Remembering Smell

Remembering
SMELL

Bonnie Blodgett

Houghton Mifflin Harcourt

BOSTON   NEW YORK

2010

For information about permission to reproduce
selections from this book, write to Permissions,
Houghton Mifflin Harcourt Publishing Company,
215 Park Avenue South, New York, New York 10003.

www.hmhbooks.com

Library of Congress Cataloging-in-Publication Data
Blodgett, Bonnie.
Remembering smell / Bonnie Blodgett.
p. cm.
Includes bibliographical references.
ISBN 978-0-618-86188-0
1. Blodgett, Bonnie — Health. 2. Smell disorders —
Patients — United States — Biography. 3. Gardeners —
United States — Biography. 4. Smell. I. Title.
RF342.B66 2010
616.8′560092 — dc22   [B]   2009046259

Book design by Melissa Lotfy

Printed in the United States of America

DOC 10 9 8 7 6 5 4 3 2 1

To Cam

The study of disease and of identity cannot be disjoined.

— OLIVER SACKS,
The Man Who Mistook His Wife for a Hat

Contents

III
TOXIC SMELL

Part I

PHANTOSMIA

1

A POWERFUL STENCH

I'M TOLD MY NOSE is my best feature. It's long and straight and has a high bridge with a bump at the top that is a perfect perch for my thick glasses. My nose is large for my face, but I have an unusually small face. That makes me thankful for my nose. No one would describe me as mousy. When I enter a room full of strangers, I can trust my nose to announce that here is a serious, thoughtful person. And by the way, where are the appetizers? Do we smell a touch of cumin?

But even though I could pick out the Chanel No. 5 from among ten other perfumes in a crowded room, there was a time when I took my sense of smell for granted. I assumed that it was indestructible. I certainly never asked myself which I valued more, my long, straight nose or what went on inside it.

My story begins on a Wisconsin interstate just before half of it veers south toward Chicago and half goes west to places you've probably never heard of—like the Wisconsin Dells, Altoona, Eau Claire—and then finally to the Twin Cities. I was driving home to St. Paul after a weekend visit with my daughter Caroline, a student at UW–Madison, when my nose began picking up a weird smell. Had I stepped in something? What could be causing this peculiar odor?

I pulled into a Kwik Trip to top off the tank and check my shoes. Nothing suspicious there. Maybe the heater fan was sucking up the smell from the engine and blowing it through the vent. Was a dead bird in there?

Ridiculous. The smell was all in my head, not my nose. Nerves. Saying goodbye to Caroline had been more difficult than usual. She was as lonely and homesick at Madison as her older sister, Alex, had been happy there. How different my girls were.

My own college years weren't exactly blissful. While other students were getting acquainted with one another, I was out foraging for plant materials, mainly tree branches of a certain shape and size, with which to transform my cinder-block cube of a dorm room into a leafy forest glade. The smells of oak leaves and pine sap soothed my homesickness for Minnesota. Years later, when my husband, Cam, and I settled down to raise a family, I couldn't wait to plant a garden. I dug up the patchy lawn in the backyard.

Gardening to me is an artistic endeavor, and a garden of one's own represents the ultimate in creative freedom. In fact, in my forties I became so greedy for that anything-goes fix I got when planning a new border or rigging up a water feature that I decided to quit my job editing a city magazine to launch a publication of my own, the *Garden Letter: Green Thoughts for the Northern Gardener*. When my little magazine won an award from the Garden Writers Association for the Art of Garden Communication, a category invented just for it, I realized I'd turned a corner: I was a garden writer.

• • •

Before leaving for Madison I'd brought in the last of the tomatoes and potted up some herbs that would spend the winter on a sunny windowsill in our kitchen; maybe that was the source of the smell. Usually when I'm driving and smell something funny, I can track it down to my fingertips. I sniffed. Rosemary. Thyme. Lemon verbena. Chili pepper. Herbs and spices have amazing staying power on the skin, but while this strange smell in my car was persistent, it wasn't anything like the delightfully pungent scents that I carried around on my hands for hours after I'd been pressed into service as Cam's sous-chef. This wasn't the smell of a garlicky pesto or a Cajun rub.

I cracked my window and sniffed the incoming air. To my nose, Wisconsin smells wonderful after the fall harvest, maybe because I grew up around farms. My favorite aroma after pine needles fermenting on a forest floor has to be baled hay laced with cow manure.

I sniffed again. For a blessed moment I thought I'd solved the riddle of the smell. But a third sniff told me unequivocally I hadn't. This wasn't cow manure. This was sickly sweet. Hog dung, maybe?

It occurred to me that I had no idea where I was or how long I'd been driving. It's easy to miss the turnoff for the Dells, the Waterpark Capital of the World. My hands tightened on the steering wheel. This little mistake can add anywhere from a couple of hours to a whole day to the trip home, depending on when you wake up and (speaking of unpleasant odors) smell the polluted air that alerts you to the fact that you're approaching a major metropolis. Interstates in the Midwest tend to look alike, the same Kwik Trips and low-slung Frank Lloyd Wright–wannabe rest areas. So it's easy to keep sailing along on the wrong ribbon of asphalt until the tickle in the back of your nose tells you that you're not headed to the Dells but to a city of 9.5 million people: Chicago.

A huge neon blue corkscrew emerged over the treetops a few miles ahead. *Please, Lord, make it be a water slide.* In this agrarian part of the state, the Dells comes on like a tsunami of kitsch. PUT YOUR FEET UP AT THE POLYNESIAN RESORT HOTEL AND SUITES — KIDS UNDER 10 STAY FREE! STOP BY GOODY GOODY GUM DROP CANDY KITCHEN. WE HAVE EVERYTHING YOUR MOUTH CAN IMAGINE! The attack of the billboards provoked my usual impulse to drive the car right through them (and maybe take out a row of those faux-log town homes and a flume ride or two). At least I was on the right highway.

The smell tiptoed back into my consciousness. What *was* it? Hot dogs? Not at this time of year. I detected a trace of dead fish. What was odd about this smell and why it *was not normal* was its refusal to back off. If anything, it seemed to be getting stronger, as if the fade button in the brain that makes an odor disappear after a while was broken. Could it be coming from the double latte I'd picked up at Starbucks on my way out of Madison? Had the milk gone bad?

I'd brought along the audio book of E. L. Doctorow's *The March,* a fictionalized account of Sherman's epic sacking of the South that ended the Civil War. Three chapters left to go. I slipped in a cassette, then decided to hold off listening until I'd gotten around the geezer in the battered pickup ahead of me.

He was driving a Chevy S-10, same model and vintage as the one I'd bought last summer for hauling brush to the municipal compost site. A twelve-year-old pickup is not ideal for freeway driving. His tailpipe was leaving quite a thick plume of exhaust. If I could just get by him, maybe the smell that had been dogging me would disappear. The pickup was in the passing lane and flanked by a late-model Buick traveling at the same speed — as if they were ballroom dancing and didn't want anyone cutting in. I bore down on the pickup until less than a yard separated his bumper from mine. The geezer finally got the hint and pulled over to the right lane. I pressed hard on the gas pedal. When the pickup had been reduced to a speck in the rearview mirror, I had to face facts: it wasn't the poisonous tailpipe. The mysterious odor hadn't budged.

My Passat was no ingénue itself. At seven years old, the car had a few bad habits, like stalling out at stoplights and turning on the check-engine light for no reason. In fact, the thing was aglow right now. The smell had a smoky and slightly chemical quality to it. I checked the heater and got the usual blast of hot air in my face but no spike in the odor's intensity. The engine sounded all right; the car was braking normally, and it wasn't vibrating the way my truck did when I pushed it over sixty-five. Plenty of gas too, so it wasn't a leak. Good. Nothing serious. The catalytic converter, maybe?

This smell had some sulfur in it for sure. It used to be that you never knew when your engine would start belching up rotten eggs. I hadn't smelled that for ages. I was tempted to call my husband, ask him if he thought a faulty converter could make a car stink. He wasn't a mechanical person, but his incredulity when I ran something like this by him was always reassuring. *You're nuts,* he'd say. Thinking you're nuts is better than thinking your car is going to blow up and send pieces of your body flying into five counties.

I pushed the button to get *The March* rolling. Listening to the descriptions of Southern towns lovely even in ruins and how the heavy perfume of a saucer magnolia had the power to bring people of different heritages and even races together took my mind off my own olfactory troubles — until a sign stuck in a tractor tire at the edge of the freeway informed me that the county I was passing through was having a festival this week and I should get off at the next exit if I wanted to participate in the Annual Firemen's Smelt Feed. The sun was sinking fast. I wondered when the smelt ran and what they smelled like and how far the smell of fried smelt could carry. Other drivers were starting to turn on their headlights. The gathering gloom slowly erased the Wisconsin hills, shrank the larger world, and made the interstate feel almost cozy, subtly altering my connection to the dwindling number of vehicles heading west.

Night pulled the shade until my solitude was complete. I couldn't wait to get home. My house, built in 1880, has been in our family for six generations. It has a pretty strong smell of its own—a blend of old carpets, dust, pets, and whatever we had for dinner last night—that would put up a fight against any intruders.

About eight miles from Appomattox, Sherman concluded his military adventure, bade farewell to his soldiers, and headed back to his family in Ohio to mourn the loss of a son killed in action just as the war was winding down. I decided not to wait for Cam's leftovers, even though he works wonders with them, and pulled into a Subway. The turkey sandwich had a peculiar taste. Was it Miracle Whip?

My husband was already asleep when I finally pulled the Passat into the garage and then crept up the back stairs to our bedroom. I fell in bed beside him, relieved not to have to tell him what had been bothering me all the way home from Madison—and was still bothering me.

The next morning, I followed Cam into the bathroom. Actually, I followed the smell. It greeted me right away when I woke up. I had an immediate impulse to open some windows, get some air in the room. Had my husband switched to a new after-shave with a lot of musk in it? No, he was still shaving. He had a towel around his waist and a wet head. His cheeks were inch-deep in shaving cream. Was *that* it?

"You're up early," he said.

"Do you smell that?" I asked, trying not to notice the odor's remarkable similarity to the smell of whatever was going on in my car the day before. "Did the cat kill another mouse?"

He smelled nothing unusual. He didn't argue with the mouse theory, though. I always smell things he doesn't, and our old house breeds mice. Our distant and secretive cat makes her presence known only when she's caught one. Attempting to track down the odor to its source would be futile. We'd just have to wait it out.

The odor was sickly sweet, and I admitted to myself why it was more than a mild annoyance. It didn't smell like hog dung, dead fish, sour milk, sulfur, smoke, or musk as much as it smelled like death. Rotting flesh. Roadkill. *Carrion* is the polite term. It smelled the way *Amorphophallus titanum* (a.k.a. the corpse flower) smells when it opens. The enormous red bloom has an odor so similar to bad meat that carrion beetles stampede to pollinate it, and humans with sensitive noses can't handle the stench. When *A. titanum* bloomed for the first time at a conservatory in St. Paul, the event was given lots of media play. The conservatory even had a webcam installed so the public could track the flower's progress. People thronged to catch a sniff. The stench was even worse than they'd expected; it really *did* smell like rotting flesh. That deceptively beautiful flower got people in the door all right, but not many of them stayed long.

The stench was bad the next day and worse the day after that. Could it be a drug interaction? I'd recently switched from

a beta-blocker to a different pill to control my blood pressure.
Maybe this new drug didn't mix with all the other stuff I was
taking—Synthroid for a thyroid condition, multivitamins, cal-
cium and Fosamax for thinning bones. It was quite a cocktail.
For the first time, I told Cam that I was scared. He offered to
phone his friend Dan, a physician.

Dan tried not to sound skeptical when Cam gave him this
report of nose trouble. Smells. Weird smells. "No, Dan, I can't
smell anything—it's just Bon." Dan said his wife smelled weird
things occasionally too—"when she has her sinus infections,
pretty awful stuff apparently."

"That makes sense."

So the smells were just the lingering effects of that head
cold I'd had three weeks ago and only vaguely remembered? I'd
been trying to write, and my nose had kept dripping on the
keyboard.

Dan prescribed an antibiotic and recommended I use a
steamer to open up my sinuses. The steamer cleared my head
all right, but it didn't take care of the smell. If anything, the
disgusting, dead-animal sweetness intensified. Bent over the
device with my nose locked onto its rubber collar, I gave in to
an eerie detachment. My glasses fogged up, and my mind with
them.

Cam was starting to get the picture. My sinuses weren't causing
this.

"What does it smell like now?" he asked.

"Imagine every disgusting thing you can think of tossed
into a blender and puréed," I finally said for lack of any other
comparison that I hadn't already banged to death. "Now take
off the lid and stick your nose in it."

The steamer's rubber collar made my voice nasal and distant. Cam said I sounded like I was talking through a garden hose. My husband has a penchant for dreaming up offbeat analogies. Hoping to cheer me up, he said he couldn't stop thinking about those long-nosed monkeys in Borneo that looked like they had tubes attached to their faces.

"Everything still stinks then, just as bad as before?"

"Do you honestly want to know?" I asked.

"Of course I want to know."

"Worse."

As the weeks passed and the smell didn't, Cam wondered what was *really* bothering me. Was it the fast-approaching holiday season? Menopause? Our marriage? Is that why I thought he smelled?

My internist too kept mining my personal life for answers. How were the girls? Was my mother still sick? When I insisted that my new blood pressure medication was causing the smell, she reluctantly switched me to a different one. She also prescribed a mild tranquilizer for my nerves. I was astonished by its cool efficiency. An invisible hand gently put my brain back in order. Anxiety was erased as completely as a set of equations off a chalkboard. I could get used to this. Hooked.

Unfortunately, the drug failed to jump-start negotiations for a lasting truce between me and "this smell thing"—which was how my doctor now referred to my condition. After I spent a few hours in la-la land (with no respite from the smell), the racing heart and roiling gut—symptoms that had become as constant as the odor itself—returned.

I wanted to keep popping those amazing little yellow pills but kept running into my inner hall monitor on the way to the medicine cabinet. Nurse Ratched on patrol. *Tranquilizers are for genuine crises,* she'd scold. *A messy divorce, or a death, or a job layoff when you have enough credit-card debt to drive you into bankruptcy. You don't take tranquilizers when all you're suffering from is some nameless odor that no one can smell but you.*

What I needed to do, I told myself, was a bit of mental tap-dancing—convince myself that the smell wasn't all that strange or scary, that it would go away, and that even if it didn't go away, I could live with it. I could tune it out as I would any other minor irritation, like low back pain, a runny nose, the pain and itch of hemorrhoids. If only living with a constant stench were a minor irritation or the sort of routine complaint you could whine about to a girlfriend whose hemorrhoids were as bad as yours.

Besides, I'd already tried "giving in" to the smell. Did I have to keep trying to give in? Yes, *try harder.*

I developed another theory: I was allergic to my wristwatch. I called my doctor to ask her if I should switch from an aluminum to a cloth watchband. She said she'd been meaning to get back to me. I had something called burning mouth syndrome. "It may be psychiatric in origin. Your mouth just goes nuts for no reason. Everything tastes bad. Like really, really hot."

She referred me to an ear, nose, and throat specialist. "He'll be able to rule out anything else." She added, "Then I'd suggest you see a psychiatrist."

2

DIAGNOSIS

M Y OLDER DAUGHTER, Alex, came along with me to the ENT doctor, whom henceforth I will refer to as Dr. Cushing. Alex lived in New York City and had asked for a few days off from her job to help me get ready for the holidays. That was her stated reason, anyway. I knew the real one: she didn't like what she was hearing from Cam.

I'd read in one of those health articles that fill the Sunday newspaper supplement that distorted smells can be an early indicator of schizophrenia. One theory to explain the connection is that schizophrenia appears to be associated with abnormalities in the hippocampus; this organ in the brain plays a major role in bringing together environmental signals and helping to form a coherent view of the world, and it's also an essential part of the olfactory system. Cam knew I was no more likely than he was to be diagnosed with schizophrenia. He also knew I'd nonetheless diagnose myself as schizophrenic if this smell didn't go away.

A short roundish man with glasses and thinning black hair greeted me and my daughter with a brisk handshake each. His manner was hurried, but he seemed nice enough: "What brings the two of you here today?" Dr. Cushing asked.

I'd rehearsed my spiel; I didn't want to leave out a single detail that might turn into a vital clue. I told him about the drive home from Madison, the various causes I'd assigned to the odor along the way: coffee, truck exhaust, my own car. I wanted him to know exactly what the smell was like and how it came in waves with various but all equally awful notes competing for dominance. I could tell he wasn't paying much attention. Why the listless response?

Anyone with my problem would know immediately what I was talking about. My verbal renderings of the smell grew more and more overwrought as it became clear — at least in my mind, which probably was not the best gauge — that this man did *not* know what I was talking about. Was he a gardener? I asked. Yes, he was. Did he grow snakeroot? No? I suggested that he might know this popular late-blooming perennial by one of its other common names, bugbane or black cohosh, or by its Latin name, which used to be *Cimicifuga racemosa* but was recently changed to *Actaea racemosa*. Either way, it stank to high heaven when in flower. Had he ever noticed?

"Sorry, I haven't," he said.

Why no questions? Shouldn't he be asking for more symptoms? I mentioned burning mouth syndrome and my aluminum watchband. He only chuckled. I told him my husband still thought it was a sinus infection. I'd had a cold a few weeks ago. I'd been using a steamer.

"Did you take anything for it?" the doctor asked.

"Just the amoxicillin. I suppose it's possible that I've built up an immunity." (How delightful to finally be conversing.) "My kids lived on that stuff. One ear infection after the other."

"I mean when the cold was coming on," he interjected. "Any over-the-counter medications?"

"I don't take cold medications," I said. Then I remembered I had taken something. A nasal spray. "Some homeopathic thing my husband bought. If anything, it made it worse."

"Zicam?"

"That was it."

"How much of it?"

Dr. Cushing was wide awake now and looking me right in the eye. How could I have omitted this important fact? Because until now it had seemed so *un*important. And so dumb. I wasn't into homeopathic remedies. A neurologist friend in California had recommended this one to Cam. Everyone out there was using the stuff. "Maybe it works, maybe it doesn't, who knows?" Cam had said. "Anyway, give it a try if you think you're getting sick. What have you got to lose?"

"How much of it went up your nose?" Dr. Cushing repeated.

I tried to recall. I went back to the beginning — that conversation with Cam. We'd been driving to the airport. He was off to New York for a trade show and was feeling cranky and tired. A bad cold had persuaded him to pick up the Zicam — for next time, he said. I'd noticed my throat felt scratchy as we pulled up to the sidewalk check-in counter.

"Damn, you *have* given me your cold," I said.

"Sorry."

"You can't get your own cold back," I joked, kissing him on the lips.

The next day I felt clobbered and useless. By the time I went to bed, the chills, headache, and raging sore throat were like a chorus of nagging crones telling me it was now or never if I wanted to slow this thing down. After an hour or so of pointless tossing from side to side, I got up and debated trying the Zicam.

Cam was right. What did I have to lose?

At first, in the dim light and without my glasses, I couldn't find the small white bottle with the orange label, so when I finally did a wave of relief swept over me and I wondered why I'd hesitated. I tilted my head back—the more the better, I figured—and sniffed hard from the pointed rubber tip, the index finger of my left hand covering my left nostril to maximize the force of my inhale. I squeezed again, even harder this time, with my head tilted farther back. A glob of gel swooshed up my nose, cold at first but then within seconds burning into my membranes. I went to work on the other nostril.

My sinuses were suddenly on fire, and so were the tissues far back in my throat where the gel had apparently started to drip. *So this is how cattle feel under a branding iron,* I thought as tears filled my eyes. One made it over the edge and rolled down my cheek. But even though my nose felt like someone was burning a match up there while pouring a steady trickle of battery acid down my throat, and I had to take a Tylenol No. 3 to get to sleep, I didn't panic. So I'd have a sore nose for a while.

"I've stopped recommending Zicam to my patients," Dr. Cushing said when I finished my story. "The nasal spray has been causing some problems. Problems like yours."

His personal opinion was that the active ingredient in Zicam Cold Remedy nasal gel spray, zinc gluconate, was toxic to smell receptor neurons, but no studies had proved this conclusively because cold infections too can harm the nasal receptors. Colds and head injuries are the leading causes of smell dysfunction, Dr. Cushing said. "Head injuries are the more serious of the two. They can permanently destroy the olfactory receptor sheet. It doesn't take much of a blow to the head to sever the nerves completely."

"I'm afraid you lost me back at *receptor neurons*," I said.

"The receptor sheet is where smelling begins. The technical term is *olfactory epithelium*, but I think *receptor sheet* gives you a better visual picture of how smell works."

He went on to explain that smell is called a chemical sense because it's triggered by substances with distinct molecular structures. The olfactory epithelium (or receptor sheet) is where proteins called receptors receive odorants—volatile airborne molecules—bind with them, and then send electrical signals up the nerve axons for interpretation by the brain's limbic system and olfactory cortex. Dr. Cushing explained that a glitch at any stage of processing could cause (a) no smell, (b) weird smells, (c) a serious brain disorder, or (d) all of the above.

It's a cliché to compare the brain to a computer, he went on, but the similarities are remarkable—except that computers don't have neurotransmitters and hormones in synapses controlling their digital impulses or DNA speaking in complex codes. Also, and critically important, computers are man-made, while the human brain is alive and still evolving and therefore, almost by definition, not entirely known or knowable.

I began to fiddle with my purse. Glitches? Serious brain disorders? Where was this headed? My eyes met my daughter's. Alex replied with a weak smile that said *You got me.*

"You were talking about head injuries," I reminded the doctor. "What about colds? Did you say a cold could damage your receptor sheet?"

Dr. Cushing nodded. "Certainly."

So why did he think Zicam and not the cold was the culprit in my case? I asked.

"I just . . . my gut tells me, based on how the gel affected you, the pain you describe, that Zicam is the cause."

Dr. Cushing decided it was time to trot out some statistics. An estimated half a million people annually see doctors about olfactory problems. Two percent of the population below sixty-five is essentially smell-blind. A quarter of these people were born without smell. Smell diminishes precipitously in old age; half of all eighty-year-olds report reduced smell.

He eventually got to some numbers I cared about. Ten percent of people who lose the sense of smell after a head injury get it back, but for those patients whose loss of smell stems from colds, the odds improve fairly dramatically—about 30 percent get it back.

"These are very ballpark estimates," he added. "Again, we don't really know."

"And the Zicam users? What about them?"

Dr. Cushing explained that the Food and Drug Administration does not regulate homeopathic remedies, and Zicam falls into that category. There's no requirement to prove the safety and efficacy of a homeopathic product, so there are no reliable data showing if it's helpful or harmful. In spite of that, he said, in 2005 Zicam's manufacturer, a company called Matrixx Initiatives, settled a class-action suit brought by Zicam victims totaling $12.5 million. "That tells me there's a problem and they know it."

No wonder homeopathic remedies are so popular. Their promoters can make promises and then escape any consequences if the remedies fail to deliver or turn out to be harmful. I found out later that the right-wing radio spin doctor Rush Limbaugh was Zicam's advertising voice, and this seemed like a match made in heaven. The homeopathic-drug makers have nothing to gain from big government.

I made a mental note to call one of my neighbors, a personal-injury attorney, when I got home. Maybe I could get in on some of that class-action cash. Maybe I could even take on this Matrixx Initiatives myself and win. Force them to stop making the stuff. I would just have to prove that the disgusting odors brought to me by the makers of Zicam were as emotionally debilitating as they, in fact, were. What jury in its right mind would choose the free-market philosophy over consumer protection in a case like this?

"So how long before I'll notice any improvement?" I asked. It hadn't sunk in that there might not be any improvement.

"Cell repair takes time, up to six months," he said.

"A CT scan will almost certainly rule out a brain tumor," he added. "Be sure and stop by the front desk to schedule one. We'll get it done later today so you'll have the results tomorrow."

He pulled his stool closer until he was sitting opposite me, knees to knees. He needed to look at my sinuses, just to make sure no polyps were distorting my smell. Unable to grasp how a common cold or, even more improbably, an over-the-counter cold remedy could so royally mess up my sense of smell, I trusted that his long metal scope, thinner than a paper clip, would find the real problem and fix it.

He strapped on a headlamp. A white light beamed from his forehead like a third eye. His nurse sat beside me and took my hand. "Try not to move," she said. "This could sting." Hearing this, Alex moved her chair closer and took my other hand. "Squeeze when it hurts," she whispered. I closed my eyes. Something cool pressed against my nostril. Then a series of sharp pricks. I squeezed Alex's hand. Hard.

"That's it," Dr. Cushing said when he was finished. "Not so bad, eh? Your sinuses are clear, no sign of polyps." He reiterated that the bad smells were nothing serious. Nothing, period. The smells weren't real. "We call them olfactory hallucinations. The patient smells things no one else can."

Then came the thunderbolt.

"I'm afraid that once the phantosmia stops, you won't be able to smell anything. You may find it a relief."

"Excuse me?"

"You've lost your sense of smell."

He handed me a box of Kleenex (his scope had given me the sniffles) and suggested that I should consider myself fortunate. He'd recently given the same bad news to a patient who made his living as a chef.

"What did he do?" I asked. Stupid question, but I had to say something, if only to avoid falling apart in this man's office.

"He had to figure out another way to make a living, of course. Did you tell me you're a writer? That's lucky for you."

I suddenly detested this man. How could I find comfort in the fact that I don't make my living "by my nose"? In fact, as a garden writer, I do. And I don't just write about plants. I enjoy smelling them almost as much as looking at them. My garden smells like, like . . . like heaven, in late summer especially.

I put the Kleenex box on the floor beside me. I was beyond crying. Phantosmia, this harbinger of an odorless existence, was, whatever this doctor might think, profoundly disturbing. How would he know what phantosmia felt like? How would anyone? People get colds. They get cancer. They don't get—

"*Anosmia* is the medical term for loss of smell," he went on. "*Osmia* means 'smell' in Greek, and *an*, of course, is the Latin word for 'without.' I'm going to put you on steroids, though they usually don't help with healing unless they're given immediately after the injury. And it's been . . . what, three months now?"

It had actually been just a little over eight weeks. Now there was no doubt in my mind—or seemingly in his—that it was Zicam that had destroyed my ability to smell. What else it might have done to my brain was anyone's guess. So why hadn't the stuff been banned? Joining the class-action suit seemed laughable now. How could a few thousand bucks compensate for this? How could I have considered joining a lawsuit, calling up my friend the lawyer? The diagnosis of a temporary olfactory malfunction was one thing, but permanent loss of smell? Unthinkable.

Dr. Cushing said he wanted to put me on an antidepressant to treat the phantom smells. "It's an old-fashioned tricyclic called amitriptyline, a very good and safe drug that was developed to treat depression. That was before the new SSRIs, which directly target the low serotonin levels that are the underlying cause of depression in most people. You've heard of Prozac. Tricyclics are a blunt instrument by comparison, and we still don't fully understand how they work on brain disorders like yours. The drug may give you a little dry mouth, and you'll have to watch your alcohol intake and sun exposure. It's all explained in the directions."

"But I'm not depressed. At least, I wasn't before this happened."

"No, what you have is a sort of psychosis. Like your brain is having a panic attack. The drug seems to trick the brain into settling down. Maybe the brain thinks you're smelling again and stops trying to do it for you. Phantom sensations are not well understood. There's a disconnect between the brain and the body. Signals get crossed. Nerves misfire. But it's possible the brain is trying to compensate for the lost or damaged body part. It'll give up after a while even without the drug, but why wait when we can shut the smell factory down with a pill?"

This was astounding. Phantosmia is the olfactory equivalent of phantom limb syndrome?

Too upset to absorb the full implications of Dr. Cushing's latest revelation, I fixed my attention instead on the antidepressant he wanted me to take to extinguish the bad smells. Why would I want to erase the last vestiges — even if they were the last gasp, the death throes — of my beloved sense of smell? I finally managed to ask a good question.

"Why are the smells so awful?"

"I'm afraid we don't know." Dr. Cushing was silent for a few seconds, mulling this over. "One could speculate, of course."

He checked his watch and rose from his stool, which clearly meant *We won't be going there. Not today.* "I want you to take the pill before bed. It will help you sleep. Insomnia can be a problem for patients with phantosmia. During the day you can take a tranquilizer if you need it." He was scribbling on his prescription pad now. "The drug is called Ativan. It's similar to Valium." This was the same pill my internist had prescribed. It was effective, all right, if you wanted a cumulus cloud for a brain. He added, "Don't be concerned if they make you feel a little anxious. The smells, I mean. It's a normal response."

Normal? How could this be in any way normal? To be fair, I *had* told him just twenty minutes earlier that deliverance from foul odors was my fondest wish. Deliver me forever from smell itself, if that's what it takes, I'd told him. I now officially emended that statement. Without smell, my long, straight nose was a joke, an impostor. My nose was a useless lump of bone and cartilage, good for only one thing: breathing. What is breathing, what is living, without smell?

"Just remember that the phantosmia you're experiencing is nothing to be concerned about," Dr. Cushing was saying. "That's *phant*, of course, from *phantom*, and—"

"—*osmia*, meaning 'smell,'" I blurted out before he could.

"Exactly. The drug will work, and even if it doesn't, the smells will stop eventually."

"When?"

"Hard to say. Sometimes it takes years. But the odds of them lasting forever are very low. Fortunately, with the amitriptyline you won't have to wait long at all. I'm going to go out on a limb and *promise* you that you won't be smelling a thing by Christmas."

"So this really is permanent?"

"You mean the anosmia?"

That word again. I nodded.

"Most likely, yes," he said. "Just in case the cells aren't all dead, I'm putting you on a five-day course of steroids, the standard treatment. Steroids can sometimes jump-start a healing process. But again, they usually don't help unless you start taking them immediately after the injury. Given that, well—I guess I wouldn't get my hopes up."

"So there's no way of knowing?"

"You could have a piece of the olfactory epithelium biopsied," he said. He didn't recommend it. "It's excruciating and dangerous. We're talking about brain cells. You don't want to mess with those if you don't have to."

Dr. Cushing extended his right hand to signal that our appointment was over. "I'll want to see you again in three weeks," he said. "You'll lose some weight at first. All my anosmics do.

You'll put it back on and probably more over the next year or so. The typical anosmic is about twenty-five pounds overweight. They think this has to do with satiety. The hormone that's released in the brain to tell us we're full can't function without a working nose.

"For now, though, you're going to have to watch the appetite. You may have to force yourself to eat."

3

NOTHING REALLY SERIOUS

ALEX GENTLY REMINDED me we had shopping to do. Though as stunned by my new diagnosis as I was, she immediately became my life support. Just by holding my hand, she allowed me to inhale and then exhale. Get out of the chair and find my purse. I was able to schedule the CT scan, walk to the elevator, and get the scan done. With Alex beside me, her hand in mine, I found the car in the parking lot and nodded in agreement when she suggested we stop at Macy's on the way home. Somehow I drove. I entered the department store and together we picked out a new sweater for Cam, a skirt for Caroline. Eventually I found my voice and began spouting platitudes: Aren't I lucky it's just my nose? Think of what others endure. People just go on. People live full, rich lives even when they're confined to wheelchairs and can't move, when they're deaf and blind—think about Helen Keller—and even when they know they're dying, for God's sake. People really are remarkable, aren't they?

By the time Alex and I got home, my mood had gone from giddy to black. When I announced to my husband that I'd never smell again, something in his manner—his struck-dumb disbelief?—made me feel as if I were being pulled out to sea by an undertow. This sensation did not fade even as we talked. I watched as Cam rolled up newspapers and stuffed them under the fireplace grate, laid down the logs, and then lit the match, blowing on the tiny bluish flame to encourage it to spread so the thin birch log on the bottom would catch. Confessing my immediate fears (what would the CT scan discover?) to my husband was out of the question. I already knew the scan was nonsense. Dr. Cushing had found nothing suspicious when he put his scope up my nose, but he had made a curious comment. The long, thin olfactory bulb connects directly to the amygdala and the septum, which are parts of the limbic system, the structure that controls our most emotional and instinctive behaviors. Many physicians consider the organ for smell part of the limbic system, and not just because the two are located close to each other. Unlike images and sounds, one pathway for odors goes directly to the brain's emotion and memory centers without being filtered by the circuits involved in higher intelligence.

Fear is a hard-wired emotion tied to survival. It is the amygdala's default setting, the primal emotion in all animals. In humans, brain areas responsible for problem-solving capabilities, such as superior vision and executive function, developed and formed in layers around the ancient midbrain, where the limbic system is located. But the midbrain maintained its primal status and situation, just upstream from the nasal passages. This means that once fear takes over, reason may be a help or a hindrance, but it will always be secondary. As Jonah Lehrer explains in his book *How We Decide*, contrary to popular (human) belief, we are a feeling species first, a thinking species second. This I knew from personal experience. Underlying my various worst-case scenarios—only yesterday I'd been fretting about schizophrenia, and now I was imagining inoperable cancer—

was the grim reality that those crazy notions had been summoned to suppress: the reality of anosmia itself. *I can't smell and I don't know what that means.* Without my sense of smell, the world would seem flat and featureless; tasteless too.

Don't go there.

Sitting by the fire with Cam, I painted an uncharitable picture of Dr. Cushing. Reasonableness and calm were impossible responses to such a catastrophic loss. The limbic system had taken full possession of my wits. Instead of overwhelming it with logical arguments, I tried to focus my overheated amygdala in a new direction, toward anger and away from fear. I put on my best scowl and raised my voice to stop its quavering. That man, I complained loudly, had been attached to his scope and headlamp so long, he seemed to think people were basically computers. He'd probably completely lose it if he found dog poop on his shoe.

"So he really thinks the Zicam zapped your nose?" Cam said.

"Are you trying to tell me you *don't?*" I snapped. "I'm just making this up? Is that what you really think?"

My husband sat beside me on the sofa. "Of course I don't think that," he said.

How I wished I could be angry, and that I could believe my husband.

In the film *Groundhog Day,* the character played by Bill Murray wakes up every day to find he's still living the previous one. Stuck in time, he is frozen in place. I was that character. *I can't smell and I don't know what that means.* The tape was my wake-up call, the shrill morning alarm that shattered the blessed relief of slumber and told the limbic system to rise and shine. Time to jolt the heart into overdrive and turn on the faucet marked ACID to begin the gastrointestinal slow burn. Sleep was now out of the question unless I was willing to drug myself with a megadose of those little yellow pills, the tranquilizer called Ativan, which I wasn't, because as bad as this was, giving in to fear would be worse.

Cam seemed to have bought my misleading portrayal of Dr. Cushing. The poor man was undoubtedly a quack. Cam ridiculed the notion that a nasal spray could destroy a person's sense of smell. That it was he who'd recommended I use Zicam cemented his conviction that the cause was not some innocuous gel but the cold itself—"if in fact you actually *have* lost your smell."

Dr. Cushing called with good news: the CT scan was clear. I did not have a brain tumor. I heard a click on the upstairs line. Cam joined the conversation. The doctor reiterated that my anosmia was almost certainly permanent. He repeated his theory that Zicam was the cause, adding that Cam shouldn't blame himself; millions of people were using the stuff, even though it had only a placebo effect at best, and loss of olfactory function was an exceedingly rare, albeit tragic, consequence. He also told Cam that "there is no test."

This was in response to my husband's suggestion that he "stick something up there and take a look." Feeling defensive on the doctor's behalf, especially in light of my unkind remarks the night before, I reminded him that Dr. Cushing already *had* taken a look.

There was a test that I could self-administer, the doctor said. The University of Pennsylvania Smell Inventory Test (UPSIT) rates the severity and specificity of smell loss. It measures a person's ability to identify up to forty different odors. Dr. Cushing said he'd found such tests unhelpful, however. "If patients tell me they can identify smells, I know they're probably lying."

While I tried to interpret this bizarre statement, Cam used it as the basis for his budding theory of a psychosomatic cause for my phantosmia. He suggested to Dr. Cushing that maybe I was just imagining all this. "She gets migraine headaches too," he told the doctor. "Emotional stress can have physical manifestations, isn't that so?"

Dr. Cushing said he left that sort of thing to the psychiatrists and maybe I should see one. All he had to go on was my own testimony that I was smelling foul odors, a classic sign of a damaged receptor sheet. Then he wished us both good luck and reminded me to come back in three weeks so he could check my "progress." He signed off on what he must have thought sounded like a positive note: "Remember what I said. You won't be smelling a thing by Christmas. The treatment usually takes about four days to start working."

Cam immediately suggested we see a psychiatrist. He said he'd come with me. If the psychiatrist believed, as Cam did, that the awful odors were psychological, then all I'd have to do was "let them go" and get on with the task of sorting out the real cause of my distress. The psychiatrist would help me. We'd have long talks. Eventually we'd talk my troubles away.

"You are obviously extremely anxious," the psychiatrist said.

Then he asked me how my mood was, apart from the anxiety, and had me count backward from one hundred by sevens. I got to ninety-three. What was going on? I used to be able to count backward by sevens.

Apparently satisfied, the psychiatrist scribbled the name of a new drug on a prescription pad, along with an Ativan refill. The drug was Lexapro, one of the SSRIs (selective serotonin reuptake inhibitors) that Dr. Cushing had told me about. The Ativan would keep my anxiety under control while the SSRI built up in my system. That could take a month or more. "Everyone reacts differently." The SSRI, if it worked, would deal with the chemical cause of my tendency to hit the panic button, as Cam often referred to that aspect of my personality. And it was not addictive. Nor would it put me into an altered state that could be habit-forming, as the yellow pills had been known to do. As for the tricyclics, he assured us that they were perfectly safe. While he hadn't heard of phantosmia, he had read some phantom-limb-syndrome case studies in medical school. Not long ago he'd treated a veteran with PLS. Awful stuff. Worse than losing the leg, the amputee had told him. "Almost as bad as the nightmares. The poor kid had done a tour in Iraq," the psychiatrist explained.

Driving home, Cam went on about "what a nice guy" and "how helpful" while my thinking brain totted up the clues and ran them by the limbic system. Then it dawned on me what had been going on in that office. The counting-backward exercise made perfect sense. The psychiatrist was testing me for early signs of Alzheimer's. Phantosmia was just the opening salvo. I was on my way to complete mental and physical disintegration.

Caroline had arranged to take a bus home from Madison for midwinter break. I tried to get some work done before picking her up at the Greyhound depot. The bills had been piling up. I noticed as I signed the first of the checks that the pen wobbled in my hand. I seemed to have acquired an elderly person's penmanship. I already had a slight tremor. It had started maybe three years ago, and it came and went like a cat. I'd decided to ignore it after my doctor guessed that I'd damaged a nerve in my neck hauling flagstones for a terrace I'd installed one spring.

But this morning the tremor was definitely more pronounced. What's more, I couldn't seem to focus my eyes. The fine print on the checks was blurry. Was my vision going now too? That ruled out Alzheimer's. I couldn't help myself. I went online and immediately hit pay dirt. Smell dysfunction and blurred vision are both early warning signs of multiple sclerosis. The disease has no known cause and often attacks women during and immediately after menopause.

Caroline wasn't hard to pick out among the students filing off the bus. Her heavy, almost waist-length brown hair had been shoved up into a bulging topknot that listed toward her left ear. She wore bright red Badger sweatpants and a black North Face parka with duct tape wrapped around the sleeve where she'd snagged it on something and opened a half-inch gash in the fabric.

Before long I was dropping hints about eye trouble.

"Do you think you should be driving, Mom?" Caroline asked. Her sense of humor was intact.

"Do you think we'll get to see you without that parka on?" I shot back.

I suspected she slept in it. Maybe it was a substitute for the stuffed monkey she'd had since she was four and left at home when she went to college. The parka usually smelled of her favorite perfume, a new woodsy scent by Ralph Lauren, and body odor. This morning, the smell sent my nose into overdrive. Just as Cam's shaving cream had that first awful morning after I returned from Madison. Just as toothpaste did, and coffee, and perfume. It seemed that the stronger and more familiar the actual odor was, the worse the surrogate my brain conjured up to take its place. Perhaps a few of my odor receptors still had some life but distorted the smells they detected, or maybe my brain was just freaking out. I couldn't tell if the smells were distortions of actual odors or complete inventions. But the overwhelming result was a huge disconnect between my brain and the outside world.

"It's probably just the tricyclic," I said in reference to my blurred vision.

"Probably?"

I'd left the door open a crack with that word. I wanted to be ready in case—well, just in case. Caroline stared out the window, exasperation writ large in her body language. Then she turned on the radio and complained noisily when it refused to cough up a tolerable tune. After a while she turned the radio off.

"So how are you really, Mom?" she asked. "Is the drug working? Still smelling things?"

She did not take her eyes off the dashboard. Children aren't supposed to play parent. I let the question hang in the air. Then I said that people were being very kind; they tried hard to understand. How do you sympathize with a person with nose trouble? It's not like seeing a blind man crossing a street with only a white cane to guide him. Or having to shout at a deaf aunt to make yourself heard. Lots of my friends told me they couldn't imagine what I was going through. At least they were honest.

I still hadn't answered my daughter's question. How was I, really?

"Are you scared, Mom?" Caroline finally asked. "I'd be scared."

"I keep thinking the doctors know something I don't," I said. I mumbled the name of today's dread disease: MS.

"Well, I think it's syphilis. Who've you been hanging out with? Some retard gardener, probably. Have you told Dad?"

Sarcasm added a dark bass note to her husky voice. We didn't speak for a couple of minutes. "Who gets phantosmia?" I finally asked. "That shrink hadn't even heard of it. My internist told me I had burning mouth syndrome and that it was psychiatric. Now this other syndrome. Is this whole phantosmia thing a cover? Does everyone think I'm a nut case?"

"Mom, *stop!*" Caroline said.

I stopped. My vision returned to normal in just twenty-four hours. So it was the amitriptyline. But I was still smelling things—horrible things—that no one else could.

4

DRAWING SMELL

I HEADED UPSTAIRS to my office when I got home. I like to draw, and I decided to make a sketch of the olfactory system. Maybe then I would understand it. I would surgically implant this nose into my head with the help of the Micron #05 black ink pen that I used to decorate the pages of the *Garden Letter*. It struck me as highly unlikely that I'd ever use the pen again for that purpose. Whatever madness once possessed me to perform such creative heroics would vanish along with my sense of smell.

Don't go there.

I managed to choke out a derisive laugh as the pen began to shake ever so slightly in my hand. Caroline was right, of course. No way did I have MS. What I did have was a passion for coffee that even phantosmia could not extinguish. Coffee makes everybody's hands shake. And my coffee is strong. So, assuming that my pen would settle down on the paper and something resembling a nose would materialize and I could make enough sense of the diagram I'd found online to transpose it onto the white page with everything in its place and accounted for (this would require patience and some awfully tiny letters), maybe *then* I'd understand what was happening to me—if I could just make the inside of my nose as real as the bony ridge between my eyes that I could see in the mirror and feel with my fingers, the nostrils wide open to the world within . . .

I pulled out a sheet of paper and went to work. It took a few attempts—and as each balled-up failure missed the wastebasket I tried not to take it as a bad omen—before I'd produced an acceptable likeness of my own profile. There was the high bridge, and the bump for my glasses. I drew small dots representing odor molecules rising from my favorite coffee mug, and I drew the yellow mucus high up in the nasal lining. This isn't the green mucus associated with colds. Olfactory mucus slows the odor molecules down and begins a sorting process that continues after the smell penetrates the skull. The mucus helps the odorant find its own designated receptor neuron in the tangle of cilia dangling from the olfactory epithelium, or receptor sheet. The cilia snatch the molecules in their fibrous clutches (imagine fine wisps of baby hair tossed about on a strong sniff, like lingerie on a clothesline). I drew some squiggles in the upper nose and then the receptors' long axons. *Axons* are what important nerves in the brain are called. The smell system's axons deliver an odor's decoded message to the high brain by way of the olfactory bulbs, one for each nostril, and the adjacent limbic system.

I gave the axons a sturdier appearance. They reminded me a little of snakes rising, not from a snake charmer's basket, but from a forest of bowling pins. The axons appeared to be attached to the tops of the pins, which were supposed to represent the receptors (magnified a trillion times) on the olfactory epithelium; the receptors are proteins that decode the odorants and then send electrical signals up the axons for the brain to read and respond to appropriately. En route from the receptor sheet to the olfactory bulb, the axons are bundled into groups, each with its own message to be delivered to (synapsed at) the correct location on the olfactory bulb, the nerve tract leading into the brain. From there, the signals, now assembled into a pattern that represents a smell, are transmitted to two places. One is the limbic system, the brain's emotion and memory center. The other is the higher, thinking brain, where they meet up with signals sent from the other sensory systems, most of which have taken a far less circuitous route to get there.

I drew half a Q-tip. This was supposed to be an olfactory bulb. The fat end faced outward. (The bulb tapers as it heads into the brain.) It was not a delicate Q-tip. The axon bundles

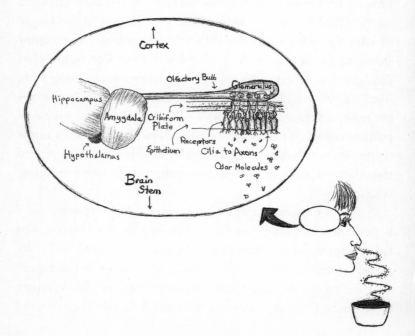

hung off the bulb like a litter of piglets sucking on a sow's teats. So much for art class.

My dog, Mel, a small, white terrier mix, was sprawled across the beanbag chair, all four legs pointing straight up. I scratched his belly and looked at my watch. Something was running around in my stomach. Hunger. Mel's belly morphed into the face of Dr. Cushing telling me, "You're going to have to watch the appetite. You may have to force yourself to eat." If an animal lost its sense of smell, would it die, since it didn't have a thinking brain to tell it sugarcoated half-truths about how life is worth living even without it? Probably. I've seen dogs in chemotherapy. It's not just nausea that puts them off food. While humans are driven by hope, longing, and dread (that is, future consequences of present acts), dogs live in the present. They do not force themselves to eat.

My questions were like odor molecules flooding a functioning receptor sheet as that lucky person sniffed what promised to be a great meal. The more we sniff, the more we can't wait to dig in. I wondered why the smell of barbecued chicken was stronger than perfume, and why perfume fades so quickly, and why some smells are pleasant and others foul (and why some are delightful to me but not my husband, or revolting to us but transcendent to our dog—I'm thinking of deer scat), and why some smells are so hard to tell apart, and why people react so differently to them.

And why we don't know.

5

AN UNDERLYING LOGIC

To understand smell, you first have to understand its primacy in human evolution. Smell is millions of years older than *Homo sapiens*, older even than man's most ancient ancestor, that nameless creature that first blundered onto land from the sea and in so doing made use of one of those evolutionary add-ons, a nose that could detect odor molecules in air as well as water.

To understand evolution, you have to erase from your brain the notion that it is a logical process resulting in exquisitely designed species.

Take the confusing jumble of cranial nerves in the human head. Paleontologist Neil Shubin likens them to the wiring and plumbing exposed when a building he once worked in was gutted. Constructed in 1896, it had undergone countless renovations. The pipes and wires, many of them damaged or useless or simply redundant, were good for something: they stored the record of how the building's mechanical infrastructure had evolved.

This I can relate to. My 1880 house still has bits of the original tube-and-post wiring, as well as quite a few ungrounded outlets and other issues that make pulling a permit a dicey proposition. The plumbing is a combination of galvanized steel, copper, and lead. We don't update the infrastructure because we'd have to tear down the house to do it. The cobbled-together system keeps us warm in winter; the appliances and lights function reasonably well; and the drinking water is as safe as the city can make it (we drink straight from the tap). My house makes a lovely metaphor for my brain. A bit of a mess, but it works.

Of the twelve paired cranial nerves in the human brain, the fifth one, the trigeminal nerve, is the most immediate example of evolution's nonlinear ways, and the most exasperating to medical students trying to keep track of its disparate functions and elaborate pathways. The trigeminal nerve controls some facial muscles and facial sensations. It registers the astringency of ammonia and the heat of jalapeños. We think of those as smells and tastes, but they're really sensations, closer to the sense of touch than to taste or smell. The trigeminal nerve also supplies the withering pain we associate with dental procedures, and the hammer blows of a migraine headache.

By contrast, the olfactory nerve is a thing of breathtaking clarity. It has one assignment only: to send odor messages from the nose to the brain.

Olfaction—the process of smelling—is another matter. It is like that building whose pipes and wires have undergone countless renovations. The human sense of smell is a work in progress whose original design has to support the weight of all the cumbersome improvements required for it to retain its value to our species.

Smelling is not only part of the limbic system—it created the limbic system. The several different species for which it was designed in the beginning depended on smell every minute of every day. Other limbic structures emerged to support this all-important alert system. The fight-or-flight response is regulated in the amygdala, while the hippocampus stores the meanings (perilous or pleasurable?) that odors convey. Deprived of smell, our dog, Mel, would be hunted down and eaten for supper by the nearest coyote; in the safety of our home, his demise would be more protracted but still inevitable without a working nose to remind him to eat. Dogs depend on smell to survive.

Humans don't. Perhaps there was a time when we did. That time is long gone. University of Chicago evolutionary biologist Yoav Gilad sparked renewed speculation about smell's long-term prospects in humans when he discovered that a certain type of rhesus monkey has color vision (most primates see in black and white). Mother Nature didn't come up with this so the monkey could enjoy the colorized version of *Citizen Kane*. Color vision enhances the monkey's ability to detect food and predators. However, it seems that the improvement required, in effect, handing over some of the rhesus monkey's smell genes in exchange for better sight. Have our human smell genes been going in the same direction?

Even if they have, smell's intimate link to the limbic system means it carries still-buried clues to how our brains evolved and to how they work now. The debate among psychologists — the few who bother to think about smell — is whether humans depend on the primary sense for emotional homeostasis. Do the myriad pleasures of scent — from the obvious ones connected with food and everyday delights like walking in the woods to the more subtle and mysterious fragrances we associate with sex — help to offset the burden that humans alone carry, the knowledge that we are mortal? Is smell's ability to trick us into losing ourselves in the moment (in pure delight) a cornerstone of human happiness? Without smell to make life worth living, indeed endurable, would our species have lasted this long?

Perfume designers call the chemicals they blend to make their alluring olfactory compositions notes. Anyone with an introductory chemistry class on her college transcript (that wouldn't be me) knows that recognizable scents are made up of molecules. The scents' formulas are like the ingredients in a recipe, a kind of shorthand that tells how much of which molecules an odor contains. The complex scent of a rose has 1,215 odor molecules; a carrot has 95.

An odor molecule must be light and lively to make it up the human nose. Most never do. Dogs smell more than we do because they have immense receptor sheets, lower stature, big snouts, and floppy ears. A bloodhound's ears are like the string mops sailors use to swab the deck: they don't let much get by them. A dog's world is saturated in smell. So is a reptile's world, but unlike snakes and lizards, dogs have excellent odor discrimination. A dog can deconstruct smell mixtures in the same way a human can inventory the contents of a room at a glance. There's the sofa, the coffee table, bookshelves, and so on. A person can't identify the ingredients in a bouillabaisse by smelling it unless he or she is a trained professional nose, as perfume designers are called. Dogs can.

So how do odorants and receptors pair up? It's critical that they do, because this hooking-up is what ensures that an odorant's electrical signal is sent to the brain. The prevailing theory is based on shape. An odor molecule and its receptor fit together like a lock and key. The odorant is the key that "unlocks" the receptor, which then sends an electrical signal along one of the nerve axons that run through the tiny holes in a wafer-thin section of the skull called the cribiform plate. Once inside the skull, the axons join together in clusters called glomeruli (pronounced "gluh-*mehr*-ya-lie" and less than expertly drawn by me) that transmit smell signals to the two olfactory bulbs located inside the brain just behind the nose and eyes and above the olfactory receptor sheet.

In the same split second that an odor molecule binds with its receptor, a signal is sent to the correct glomerulus in the olfactory bulb. A team led by Howard Hughes Medical Institute investigator Lawrence Katz of Duke University found that each glomerulus detects individual odorants only, and the olfactory bulb passes these pieces of information on to more advanced brain regions to make readable maps of the whole. The brain has to listen to each musician's melody to hear a symphony, explained Da Yu Lin, who took over the project in 2005 after Katz's death. "The whole is the sum of its parts."

Or is it? Peter Mombaerts of Rockefeller University collaborated with researchers at Yale to engineer mice that lacked a certain protein cell in the glomeruli of the olfactory bulb. The researchers don't know how the protein works, but without it, mice can't tell odors apart; smells sent on to the higher brain don't make sense. This suggests the olfactory bulb has a sorting role.

Next question: how does the brain combine sensory patterns with relevant memories, feelings, and thoughts into a single experience? In the olfactory (or piriform) cortex, where smells are consciously perceived, the odor is assigned certain characteristics specific to the smeller, such as whether or not he likes it; whether a lemon smells sour or a caramel roll sweet; and whether a particular coffee roast is pleasantly bitter or acrid.

How does the product of olfaction—this odor map created in the olfactory cortex—collaborate with inputs from other brain regions and result in (for example) someone lifting a mug of hot Kenyan to her lips, sniffing it once or twice, blowing on the surface, and sipping? What is the underlying logic of smell's passage from the receptor sheet to the neocortex, the thinking brain?

6

THE BREAKTHROUGH

THE GENES THAT CODE for olfactory receptors are the air-traffic controllers of smell. The modern conception of genes began in 1953, when James Watson and Francis Crick discovered the chemical structure of DNA, a double helix; the formation explained how genetic instructions could be stored and passed on from one generation to the next. Crick and Watson used the first image of DNA to propose what has come to be called the central dogma of molecular biology: in a nutshell, genes can make proteins (which are the building blocks of living organisms), but proteins can't make genes. Life is a one-way street. Though olfactory genes are present in every cell in the body, the genes' sole purpose is to allow us to smell. In olfactory cells, the genes are turned on. In other cells, they are not. Without smell genes to guide odorant and receptor binding, the smell brain wouldn't be able to tell a rose from a rotten egg.

How we recognize odors and how odorants bind to receptors has been a focus of the work of Columbia University geneticist and smell researcher Richard Axel. He's also concerned with what he refers to as the binding problem; as he put it in a lecture at Columbia in 2004, "How are bits of electrical activity integrated to allow for meaningful recognition of a sensory image?" How does the brain take a variety of sensory inputs and bind them together to form a full and complete perception of any one thing?

Richard Axel has long been intrigued by the binding problem in the context of the olfactory system. Unlike smell, the other senses take a direct route to the high brain. They don't pick up input from memory and emotion first, as smells do. Even taste and touch have proven relatively easy to understand, mainly because they're hard-wired. The taste of sugar remains sweet regardless of whether or not you were having a bad day when you first tasted it. But ask two people to sniff a cup of coffee and there's no telling what each will perceive. One might love the smell but call it tea; the other might know it's coffee all right but recoil in fear owing to a bad experience with burning hot coffee as a child.

Before Axel could begin to address this issue, though, he had to find the smell genes encoding odorant receptors. He assumed (correctly, as it turned out) that each gene coded for a specific smell receptor, and that each smell receptor opened the door of the olfactory system for just one particular odorant.

In 1988, Linda Buck, a sixth-year postdoctoral fellow in Axel's lab, came up with a way to identify the large family of genes encoding G protein–coupled smell receptors in the rat olfactory epithelium. These proteins tell enzymes inside the cell how to respond to an odorant. Based on what the receptor proteins *should* look like according to their genetic job description, Buck created a sort of smell-receptor-gene template consisting of three characteristics: (1) the genes that expressed the odorant-receptor proteins had to be active *only* in the olfactory epithelium; (2) the genes had to be abundant, because there were hundreds of thousands of individual odorants out there, each expecting to be greeted by a party of one; and (3) the genes had to code for proteins with a specific molecular structure that enabled them to deliver information across a cell.

Buck volunteered to put her smell-receptor-gene search engine to work on actually finding the genes. This meant she had to go through reams of lab data on mouse DNA. (The mapping of the human genome has allowed scientists to compare the human genetic blueprint with other creatures'. What separates man from mouse is minuscule, and the sense of smell isn't one of the separators.) Buck had to do her gene searching after hours; she knew that isolating the genes for the odorant receptors using all three parameters simultaneously would be tedious and time-consuming, but it was also an irresistible shortcut—and Axel didn't like shortcuts. A die-hard reductionist and devotee of the pure scientific method—he lived by the Austrian philosopher Karl Popper's doctrine that knowledge should be acquired through a process of verifying or falsifying hypotheses—Axel told Buck to take her project home.

Buck found the genes on a Saturday night. They coded for proteins with the necessary loops, all where they should be. She immediately told her boss the good news. The group of genes she'd teased out of the mouse DNA proved to be huge enough to make receptors for that warm one-on-one welcome for each odor molecule.

She and Axel coauthored the paper describing the process and outcome. Within six days of the paper's submission, *Cell* agreed to publish it; functional proof that these genes were the ones that encoded odorant receptors was delivered seven years after publication. The *Cell* article has been cited more than two thousand times in science papers. In 2004, Axel and Buck won the Nobel Prize in Physiology or Medicine for their discovery.

Olfaction may be old, but it's hardly rudimentary. There's little doubt among scientists that a thorough understanding of the molecular biology of the primary sense will lead to breathtaking new insights about not only how we smell but how we think. But scientists aren't there yet.

The discovery of such a large gene class dedicated to smell proved what evolutionary biologists had long suspected. As paleontologist Neil Shubin wrote, "Our sense of smell contains a deep record of our history as fish, amphibians, and mammals." Buck and Axel's finding was "a major breakthrough in understanding this."

In his Nobel acceptance speech, Axel called smell "the primal sense." The award alerted the world to olfaction's rock-star status in genetics, cell biology, and neuroscience. Axel made it clear, however, that the physiology of smell had not been entirely figured out. The shape theory of binding—the idea that a specific odorant fits into a specific receptor like a key in a lock, first proposed by biologist John Amoore in the 1950s—is still being debated. Cracking how the sensory inputs bind—binding problem number two—will require a breakthrough theory of consciousness. Francis Crick, of double-helix fame, was hard at work on it right up to his death in 2007. Crick and Watson's discovery of the structure of DNA came not so much from bottom-up reductionist research like Axel's but from connecting a speculative hunch with the concrete evidence in a blurry photograph of an actual double helix supplied by chemist Rosalind Franklin (see Matt Ridley's excellent biography *Francis Crick: Discoverer of the Genetic Code* for a detailed account of the matter). Their method is more akin to the work being done today in neuroscience labs that use functional MRIs and other imaging techniques to arrive at conclusions through a top-down approach. The members of Crick's team of brain scientists and geneticists at Caltech continue to pursue their late mentor's dream. Axel jokingly calls them "the ghost busters," a reference to consciousness as "the ghost in the machine."

A theory of consciousness may seem far removed from an explanation of smell dysfunction, but what if the dysfunction fills one's brain with fake smells? What do these sensations say about the nature of reality? Are they sensations at all? Could the smells be perceptions, meaning that they originate in the mind, not at the periphery, where the damage started? Scientists have shown that tinnitus, a persistent ringing, buzzing, or other sound in the ears that occurs in the absence of any external stimulus, can be a phantom sensation. They're trying to figure out how the peripheral machinery—those tiny bones in the ear—and the high brain collaborate to produce this symptom.

Tinnitus can be worse than annoying. For people who constantly hear cats screeching, cars honking, or bombs going off, tinnitus can even lead to suicide. Whatever zapped my sense of smell—and I was convinced that it was Zicam and not a cold—seemed to have caused a chain reaction. The limbic system, which responds to odors instantly, added to the confusion by drenching in fear that mother of all brain maps, the one that blends input from all brain regions to create a single conscious perception.

7

PHANTOMS

D<small>R. CUSHING'S COMPARISON</small> of phantosmia to phantom limb syndrome—"the brain is trying to compensate for the lost or damaged body part"—had me wondering if that was literally true.

Johannes Frasnelli treats smell-impaired patients in Germany. He reported that more than 60 percent of his patients with smell dysfunction had smell hallucinations, and many of these patients also suffered from depression. His American colleague Don Leopold advises ENTs on how to treat smell dysfunction. Leopold urges doctors to pay close attention when patients mention weird odors. This isn't so easy. Only 5 to 10 percent of phantosmia victims admit to having these sensations. This is because "symptoms were not taken seriously by their general practitioner [or] their families." In other words, the patient assumes he's either nuts or acting like a baby.

"Among smell pathologies," wrote psychologist and fragrance consultant Avery Gilbert in his book *What the Nose Knows*, "the most appalling is cacosmia, in which everything smells like shit." My world smelled like shit, puke, burning flesh, and rotten eggs. Not to mention smoke, chemicals, urine, and mold. My brain had truly outdone itself. If it was acting on the theory that, as with childhood fears, it was best to let the bad stuff out of the dark closet and stare it in the face, I could only offer my congratulations.

Cacosmia belongs to the phantosmia family of smells; that is, smells that have no outside source. *Parosmia* refers to bizarre distortions of *actual* odors. The smell molecules are there, but their chemical formulas are misinterpreted somewhere between the receptor sheet and the central brain. It's not always clear when phantosmia is really parosmia, or vice versa. Neuroscientist Johannes Frasnelli thinks parosmia is likely a peripheral malfunction while phantosmia denotes a central problem, such as schizophrenia or epilepsy. Phantosmia victims often report a single odor, while parosmia offers a more varied repertoire. My own symptoms put me in the parosmia camp, but I'd been diagnosed with phantosmia. You see the problem. No one really knew what I had.

With both disorders, the smells may be interpreted as familiar scents until the victim understands that the smells are not real. The fact is, the smells aren't identical to anything. That they're novel may be why people find them noxious. Novelty is always off-putting to the brain. Threatening.

Specific anosmia is the label applied to the syndrome in which a person can't smell a specific thing, such as urine, leather, or musk, but other odors come through loud and clear. *Hyposmia* means weakened smell, usually caused by the aging process. The opposite (extremely rare) condition is called *hyperosmia*. It was the subject of a chapter in Oliver Sacks's *The Man Who Mistook His Wife for a Hat*. A medical student blamed his bout of heightened smell on a drug he'd used to keep himself awake. The affliction made him able to smell as keenly as dogs, he said, and he found himself behaving like a dog, following his nose instead of his eyes and thinking brain. While this was a nuisance, he enjoyed his supercharged smeller immensely while it lasted. Oh, to be a dog again!

I continued to think of my own hallucinations as a kind of phantom limb syndrome (PLS) of the nasal cavity. My nose and brain were trying to compensate for my loss of smell function, as Dr. Cushing had suggested. My brain was becoming hysterical because my olfactory alert system had gone on the blink. *I can't find you and I need you!* And by the way, *don't touch that food!*

PLS is a brain anomaly—an exception to the rule. Thousands of amputees in the two world wars suffered phantom limb pain. Their symptoms were attributed to shell shock. Even today, unrelated symptoms—phantom pains and emotional distress—are often regarded as being part of the same thing: mental illness. We now call shell shock posttraumatic stress disorder, a more dignified and suitable term.

The power of smell is being harnessed to help soldiers recover from PTSD. A California psychologist who was deeply moved by the experiences of Iraq war veterans haunted by things they saw or did in the confusion of a guerrilla war devised a treatment. Victims were repeatedly exposed in safe surroundings to olfactory memory triggers such as the smell of burning flesh, diesel, and gunpowder, and the odor images eventually came unstuck from the memories. Eventually the smells lost their potency. They were unable to summon the emotional havoc on their own.

Traumatized amputees who come home from war sometimes attribute their PTSD symptoms not to their wartime experiences but to the phantom pains they can't understand or remedy. They are fortunate to have doctors who know that phantom sensations are real, and awful, entirely capable of making one feel crazed and out of control. PLS symptoms range from an unsettling sense that the detached limb is still there to searing sensations similar to electric shocks. Oliver Sacks described a diabetic amputee who, bedeviled by phantom sensations, complained that his doctors should have cut the nerves to the leg, then put the leg in a cast, and then, "when the feeling wasn't there, *then* cut it off! Get rid of the feeling, get rid of the idea, *then* get rid of the thing itself!" Surely this would have prevented "this damn phantom."

The man was right.

Oliver Sacks has covered the gamut of what he calls disorders of mental imaging. He had a patient who was so devastated when he lost his sense of smell after a fall that he willed it back. This belief that he could smell again—part conscious, part unconscious—intensified with time. Sacks wrote in *The Man Who Mistook His Wife for a Hat* that "he snuffs and smells the 'spring,' [calling up] a smell-memory or smell-picture so intense that he can almost deceive himself, and deceive others, into believing that he truly smells it."

Like this man's made-up smells, the odor in my brain was fake. But its origins had more in common with visual agnosia, the subject of that book's title story about the wife-turned-hat. This patient's brain sent occasional faulty pictures to his visual cortex. In the doctor's office, he tried to put his wife's head on his head, thinking it was his hat.

This question of what is real and what isn't was the topic of Sacks's third book, *A Leg to Stand On*. I read the whole book in one sitting. It tells how Sacks himself lost all sense of "connection" to a leg after an injury damaged its muscle and nerve tissue. This confounding experience threw all his assumptions about perception into the ditch. He became anxious and then morbidly depressed, and he turned to philosophy to anchor him and to poetry and music to restore his emotional equilibrium and faith in a beautiful, knowable, and reliable reality. He decided that the German philosopher Immanuel Kant had it right: human reality exists only to the extent that it conforms to the brain's ability to observe cause and effect; anything beyond that, such as who or what made us, is beyond our ken.

Later Sacks reversed himself. In a lengthy footnote attached to *Leg*'s third printing, he told of having read new theories based in part on revelations about brain function. He'd been influenced by the philosopher Thomas Kuhn, who believed that "facts are not like pebbles waiting to be picked up on the beach." Instead facts become facts based on "the viewer's theoretical viewpoint."

Well before Kuhn, the Irish philosopher George Berkeley wrote in 1710, "Light and colours, heat and cold, extension and figures—in a word, the things we think and feel—what are they but so many sensations, notions, ideas?"

What has lately been nudging scientists as well as philosophers in Berkeley's direction is the recognition that human perception presents each of us with a world more complex and varied than a mere merging of neuron patterns created in the sensory system could make possible. As abundant as receptor proteins are, the stimuli they gather cannot, on their own, create experience. The step that philosophers call binding (and that Richard Axel calls binding problem number two) must involve inputs from other brain regions that store memory and emotion. It's only when an outside event such as the loss of smell or of a limb disturbs this central binding—Axel puts all such anomalies under the umbrella category of "nonbinding"—that one gets a glimpse of how subjective and illusory reality actually is.

Experiments done in European labs found that psychosis could be induced in a subject by temporarily confusing his visual cortex with optical illusions (for example, researchers placed mirrors in such a way that they created the illusion the subject was somewhere he was in fact not). Brain scientist Vilayanur S. Ramachandran of the University of California at San Diego wondered if the technique could also be used to trick the brain into believing all was well.

Sure enough, it worked. An amputee suffering from phantom limb pain was placed in front of a mirror that reflected back his image but with his missing limb restored. Unable to tell the reflected image from the real thing, his brain assumed the missing arm was now back and stopped pestering it. Mirrors have been effective in treating Iraq war veterans' PLS. One young soldier was driving a Humvee when a roadside bomb blew off his right leg above the knee. Soon after surgery, he was harassed by phantom sensations he likened to electric shocks. Narcotics did not relieve the pain. Nothing did. His doctors persuaded the soldier to join an experiment at Walter Reed Army Medical Center in Washington, D.C., that used mirror techniques. Reluctantly, he agreed. In the study, twenty-two amputees who had each lost a leg or a foot and were experiencing phantom limb pain were divided into three groups. Each member of the first group viewed a mirror in which he saw a reflected image of his intact limb; members of the second group viewed a covered mirror; and those in the third group were

trained in mental imaging. After four weeks, all the mirror-
therapy patients reported a significant decrease in symptoms
(though some experienced intense feelings of grief on looking
in the mirror and seeing themselves as they used to be). The
soldier who'd lost his leg to the roadside bomb was in the first
group, and his pain was gone after just a few sessions.

Neuroscientists now know where his pains were coming
from. They were concocted from living memories and delivered via the nerve endings in the leg stump. When the brain
tried to move the absent limb, the result was an abnormal neural pattern experienced as searing pain. But why are phantom
feelings so nasty? Is it because the senses—touch and smell,
among others—are the first line of defense against the outside
world? The default response to a sense's absence is negative
(sharp pains, foul smells) because no other logical explanation
is available.

Writer and physician Atul Gawande describes a woman
who had spent eleven years scratching a constant itch on the
left side of her scalp until she scratched right through her skull
and into her brain and its fluid began to drip out of the wound.
The itch had started with a case of shingles that left the scalp
numb. The more she scratched, the worse the itch became. After treating her with everything from creams to tranquilizers to
brain surgery that severed nerves to the scalp, her doctors concluded that the itch was a central problem. The woman's brain
was compensating for the numb scalp, replacing the absent
sensations with sensations that felt, as Gawande put it, like
"armies of ants" crawling all over the spot. Like olfactory hallucinations, the itch is not only torturous; it's constant. Gawande's
profile of the woman tormented by this inner itch concludes
with neuroscientist Ramachandran figuratively scratching his

own scalp as he ponders what possible mirror techniques could produce the elaborate visual lie needed to finally turn off the woman's itch switch.

My phantom-smelling brain would have to be tricked — *trick* was the exact word Dr. Cushing used — into thinking all was well, just like the brains of the patients in Dr. Ramachandran's mirror experiment. But amitriptyline is a drug, not a mirror. And for all its magic-trick aspects, mirror therapy is fairly straightforward. There's still no explanation for why an antidepressant quiets the olfactory din of phantosmia beyond this: the drug just happens to bind with the receptors in the high brain that control the behavior of the misfiring synapses.

Neuroscientist Antonio Damasio explained phantom sensations as secondary brain maps made by "reconstruction through the process of recall of a previously acquired memory that kicks in when the primary maps go offline." This only applies to hearing, vision, and touch, all of which depend on a sense of the body to work properly. Smell is fundamentally different, and its connections are much more broadly dispersed in the brain because of the olfactory system's great age. Richard Axel is not alone in believing that smell is the primary sense; it has influenced human development in a profound way. But how (and why) does it continue to wield power and influence?

I asked Johannes Frasnelli if he thought phantosmia and phantom limb syndrome had anything beyond phantom sensations in common, if my mind (not the receptor sheet) could be deliberately distorting reality so that I'd smell warning odors, such as rotten food and burning flesh. I told him about my experience and asked how he explained phantosmia to his patients. Phantosmia is uncanny, I added.

Frasnelli attributes phantosmia to the brain's hard-wired aversion to novelty—in this case, unfamiliar smells caused by a malfunction of some olfactory neurons resulted in an incomplete picture of the odor formed in the olfactory bulb. Since new odors are usually perceived as unpleasant, he explained, at least in the beginning (think of exotic foods we have to get used to before we can enjoy them), this incomplete picture leads to the perception of an unpleasant smell. Frasnelli seemed to leave the door open to my rather fanciful speculation about how the brain might fall back on evolutionary imperatives and invent toxic-smelling odors as warnings.

I decided to find out if neurobiologist Donald Wilson had a theory about phantom smells. Wilson thinks all odor perception is learned (humans aren't hard-wired to dislike the smell of dead fish, for example; negative experience with dead fish gives the odor that noxious spin). In *Learning to Smell*, Wilson and his coauthor, psychologist Richard Stevenson, describe odors as highly synthetic percepts whose key components—answers to the questions "What is that smell?" and "Do I like it?"—are created in the higher brain and are dependent on conditioning.

I asked Wilson to comment on the trend among neuroscientists to challenge the preeminence of the thinking brain. (Antonio Damasio called the assumption that "the many strands of sensory processing experienced in the mind—sight and sounds, taste and aroma, surface texture and shape—all 'happen' in a single brain structure" a "false intuition.") Wilson replied with a primer on parallel processing. The neurologist Joseph LeDoux, he pointed out, has shown that when you hear a sound, that information travels to the neocortex ("reason") and at the same time to the limbic system ("emotion"). How you respond reflects a balance between those two.

On to phantosmia. Did he have any thoughts? Wilson acknowledged that phantom smells tend to be "terrible." But he went further than Frasnelli by reminding me that this is true for all phantom sensations. Wilson refrained from offering a reductive theory of phantosmia that left the other senses and the higher brain regions on the sidelines. He even tossed me a bone to chew on. The evolutionary bone.

"Maybe it's more important/adaptive for our sensory systems to provide warning than reward?"

Part II

ANOSMIA

8

THE SCENTLESS DESCENT

ONE MORNING A FEW DAYS before Christmas, I climbed the stairs to my third-floor office and fiddled with the thermostat until the furnace kicked in. Then I turned on my computer and went back down to the kitchen to fill a saucepan with water for tea. The weekly gardening column I wrote for the St. Paul paper was due at 4 P.M. It's hard to write about roses when you're engulfed in stench. I'd been putting it off. But now it was crunch time. I headed back upstairs.

The slamming of doors and the thump-thump-thumping of feet on the staircase startled me back to reality. Cam? Hadn't he left for work yet? He seemed to be taking the stairs two at a time. He appeared in my office doorway. He shoved a smoking saucepan under my nose. "Jesus Christ!" he shouted. I could see that the copper bottom was mostly black—charred, by the look of it—as was the plastic-covered handle. "What do you think you're doing?" Cam asked. (What I was doing was, obviously, not thinking.) "Thank God I forgot my BlackBerry."

He'd come back home to get it and found the kitchen filled with smoke. He stood there in my office holding a dishtowel to his face to block out the "truly hellacious odor," not yet remembering that, for all I knew, the suffocating stench was jasmine. When that dawned on him, he let fly with another "Jesus Christ," followed by "You're going to have to be more careful. You've stunk up the whole damned house."

He set the pan on a metal file cabinet and launched into a safety lecture. His voice was biting and accusatory though muffled by a second dishtowel, which had been relieved of its original purpose, protecting his hand from the pan's smoldering handle. The first dishtowel he'd placed on the file cabinet to protect the finish. Even in extremis, my husband is a careful person. Gas leaks, fires escaping the fireplace, the dire consequences of my forgetting to change the kitty litter box were mentioned. Dr. Cushing had been over most of this.

"Okay, I could care less about the damned kitty litter," Cam said, reading my thoughts, as usual. "How would you feel if the house burned down?"

I'd be dead, I almost said but didn't. He was going to have "those worthless smoke alarms" replaced. In the meantime, I *must* remember that I did not have a working nose and couldn't fall back on his when he was away. My "total self-absorption," as he put it, could have disastrous effects on others.

None of it seemed real. Still, he'd touched a nerve: others. "Caroline needs you to be strong," he'd said only the day before. Now I followed my husband downstairs, watched him carry the smoking cauldron out to the trash on his way to work. When he'd been gone five minutes I put another pan of water on the burner. As I waited for the first tiny bubbles to break on the surface, I briefly considered disabling the ignition switches on the four burners, turning on the gas, and asphyxiating myself. Everyone gets colds, suffers indigestion, strains his back, stubs her toe. Not everyone, in fact not a single person I knew . . . and so on . . . and so on. And then the tape: *I can't smell and I don't know what that means.*

When clinically depressed people say they can't think straight or read or watch TV or make sense out of the simplest recipe, this is what they're talking about. The mind of a depressed person is like a war zone. Not just frightening but very, very noisy. I couldn't shake my irrational thoughts—thought-feelings. I was stuck on the idea that I was slowly but surely coming unglued as I lost my connection to the things I cared so deeply about. The smelly things. Everything. (What *doesn't* have its own distinct fragrance?) Eventually I'd forget who I was. If psychosis didn't get me, clinical depression would.

I finished my gardening column and sent it off, knowing I'd managed to make a minor chore (putting roses to bed for the winter) sound like a major bore. *Why would any sane person do this?* was the subtext of my grim recital of dos and don'ts. *Who could possibly care?*

After a few minutes I realized I was dwelling on something even grimmer than a grim recital. I could no longer keep my mind off the countdown: if Dr. Cushing was right, I had very few days left before my ability to smell was obliterated.

In a questionnaire study, a psychologist asked his students to rank the loss of various physical traits. Losing one's sense of smell ranked somewhere at the level of losing one's big toe, with most students opting to keep the toe. Presumably the students' first thought was *How nice not to have to smell poop and vomit and stinky fish,* and it occurred to them only later that the stench coming off a visually perfect salmon fillet might save them from being poisoned and that they'd lose good smells too.

But maybe they'd still keep the toe and lose the sense of smell even after considering all that. The American Medical Association gave smell and taste a pitiful 1 to 5 percent rating as a measure of importance to a person's quality of life. The exact wording in the AMA's *Guides to the Evaluation of Permanent Impairment* is "value of a life's worth"; these ratings dictate damage awards in personal-injury cases.

Usually workers' compensation is at issue. How productive can a person be without a working nose, or leg, or right eardrum? A visual impairment is considered an 85 percent whole-person impairment (WPI); nasal dysfunction might get a 5 percent WPI award for someone who depends on her nose to make a living—a chef, for example. Someone whose career doesn't depend on smell—a hairdresser, say—might get a 1 percent WPI award for the same disability. As a garden writer, I might eke out a 2 to 3 percent impairment award.

If a photographer went blind, the damages paid would be significant, presumably because the impairment would also severely limit his options when it came to choosing a new career. The photographer might be an exceptional home cook but would not do well as a professional chef if he couldn't read a menu or navigate in a crowded and busy restaurant. The chef who became anosmic, however, could turn his amateur interest in photography into a job.

Pain and suffering don't count for anything when it comes to smell loss, even though studies on smell dysfunction and mood show that depression goes with the territory when the nose doesn't work.

Becky Phillips lost her sense of smell at about the same time I did, although Phillips never did find out why she nonetheless kept smelling "deadly urine" odors. This description convinced me that she was experiencing phantosmia—what else produces such an improbable pairing of words? Because her phantoms, unlike mine, were fleeting, they repeatedly kindled and then quashed Phillips's hopes that her sense of smell was coming back.

By her own testimony, she was a different person before she suffered the head injury that robbed her of smell. She could not go back to the life and career she'd had. It wasn't the accident but the anosmia that changed her. Before the accident, she had been super-stylish and always on the go, a successful advertising executive famous for her elaborate dinner parties and for the presentations at work that also displayed her love of sensory pleasures. "I was called the queen of ambiance," she said. "Scented candles, incense, aromatic lotions, the whole bit. I loved my sense of smell."

Phillips had been heading to an evening business event at a nightclub when a sheet of black ice sent her feet into the air and her head onto the pavement.

She'd worn her new rubber flats (with her three-inch heels tucked inside her purse) as a precaution, but they hadn't helped. Just-below-freezing temperatures and a drippy pipe above the sidewalk created a slick patch in front of the club entrance.

She told me that she fell backward and hit the ice so hard that the front of her brain banged against the inside of the front of her skull. That's how her doctors described the accident, which she can't remember. The soft tissues just behind the eyes, nose, and forehead received the worst of the blow. Not only the olfactory nerves but also the connections that help people think straight were severed.

Phillips came to in a hospital room filled with flowers. She'd been unconscious for days. Slowly Phillips remembered who she had been before the fall: the queen of ambiance. She wondered how the supposed queen could be surrounded by all these beautiful flowers and not smell them. She asked a nurse to hand her a rose, and she took a sniff. *Odd,* she thought to herself. *I can't smell it.* This realization, more than any of her other travails, sent her on a downward spiral. It took months for Phillips to accept the fact that her olfactory system was not only smashed to pieces but could not be glued back together again. Recovery was unlikely because the damage had been so complete. There were no nerve cells left alive to regenerate. She also suffered temporary amnesia, and her short-term memory was shot. She still has trouble reading.

Throughout her ordeal the single hardest part was the emotional distance she felt from people who "just don't get it" when it came to her loss of smell. She discovered, as I did, that friends did not even remember that she couldn't smell, much less understand the staggering loss this represented. They would express sympathy for her in a way that implied that the anosmia was no more significant in the grand scheme of what she'd gone through than her occasional mild headaches. Phillips had had breast cancer as a young woman, and when her cancer came back after a ten-year remission, just as she was recovering from the brutal fall, the new crisis made the disparity between what other people considered important and what she considered important all the more striking. Cancer, Phillips said, horrifies people. But telling a friend you have cancer when you're both forty gets a whole different response than it does when you're both fifty-five and know that many women have gone through it and survived. "They're not so frightened of what your cancer could mean to them. They know."

They didn't know about smell.

"All I can say, and I say this to you only because you've been there, is that I would wear my head in a bag for the rest of my life if that would give me back my sense of smell. I would pay a million dollars for what the AMA point system decided was only a nine-thousand-dollar impairment in my case. My lawyers still don't understand why [losing] something that they can't see is more important to my value as a fully functional human being than losing an arm or leg."

As Becky Phillips found out, a person who suddenly loses her sense of smell is thrown into an emotional crisis even more crippling in its way and more threatening than the loss of a leg. Why? Legs the thinking brain can get its arms around. Smell is invisible, unknowable in any concrete way. Studies have shown that even sudden blindness is less traumatic, in the long term, than anosmia, which almost always leads to depression and sometimes to suicide. This fact eluded the medical community until very recently, when revelations about the basic operation of smell began to raise troubling questions about its close proximity to the limbic system. Only now are serious studies being undertaken to examine links between smell and mental health.

Psychologist Rachel Herz tells the story of Michael Hutchence, the iconic lead singer for the Australian rock band INXS, who was found dead several years after an accident that left him a changed man and, not incidentally, an anosmic. She is convinced he committed suicide as a result of the loss of his sense of smell. His emotional breakdown—the mood swings, the irascibility, the deep and debilitating feeling of distance from the world—began when a bicycle accident (he was not wearing a helmet) left his smell receptors adrift and lifeless in the upper nose. He didn't find out until weeks after being discharged from the hospital that the olfactory nerves had been sheared by the blow to his head.

A self-described hedonist, he'd been a fun-loving person with scores of friends before the accident. He was a hard-core womanizer and a passionate gourmand. Herz believes his death was the result of the disappearance of all that after he lost his sense of smell. Hutchence described his descent into melancholy as a kind of psychosis, a profound isolation, distance, unfamiliarity with a world he'd so greedily enjoyed when he could smell it. Even blindness would not have had such a pernicious effect on him. That his despair was not rational—it made no sense even to him—of course exacerbated the depression. None of it made sense. Had he lost his vision or hearing, the impairment would have changed him forever, he knew that. But that knowledge would have saved him from his otherworldly hell. Reason is a powerful therapeutic force. It gives a person tools, a plan, a project, which can lead to a sense of pride and accomplishment. Gone blind? Learn Braille. Practice walking with a white cane. Get a Seeing Eye dog. Harness the empathy of those around you and enhance your awareness of your other senses: touch, hearing, and smell.

To support her theory of Hutchence's death, Herz points to research showing that a species' ability to sense odor nuance is in inverse proportion to the species' odor acuity. Whether this implies an emotional component, an intellectual one, or both is unclear. But humans' unique gift of flavor perception, a highly sophisticated process, undoubtedly grew out of the same evolutionary forces as smell. Even without a connection to a high brain, an animal suffers when it can't smell. Herz tells of lab experiments in which rats whose olfactory bulbs had been surgically removed became listless and refused to eat. They seemed indifferent to toys and companions. They displayed "behavioral and physiological changes that are strikingly similar to those that occur in depressed people." As a witness for plaintiffs who seek damages for emotional distress due to smell dysfunction, Herz has met enough human anosmics to have become a passionate advocate for the smelling-impaired. She is their angel of mercy. Anosmia isn't the sort of disorder one would dream up to scam an insurance company. Much better to fake whiplash or slice off a toe.

Herz wrote that "the neurological interconnection between the sense of smell (olfaction) and emotion is uniquely intimate. The areas of the brain that process smell and emotion are as intertwined and codependent as any two regions in the brain could possibly be."

Rachel Herz's mentor at Brown University was the psychologist and smell expert Trygg Engen. He knew that just because smell operates subliminally did not mean that it lacked influence. A perceptive and compassionate scientist, Engen regarded any sort of smell trouble as potentially catastrophic to one's quality of life, because the sense of smell was like an idiot savant:

> It is very sensitive, learns quickly, and forgets nothing, but it has no judgment about what ought to be remembered and what might as well be forgotten. This modus operandi will lead to many mistakes and false alarms. However, it ensures the identification of odors vital to the individual's physical and psychological well-being.

Among the beneficiaries of Trygg Engen's work and that of his successors, psychologists like Herz as well as cell biologists and other scientists newly interested in the primal sense, are people with brain disorders that alter their reality so profoundly they find themselves sinking into madness. Members of this group include schizophrenics, for whom smell dysfunction is a characteristic symptom.

Elizabeth Zierah had been anosmic for three years when, in 2008, she decided to talk about it online. A victim of a stroke in her early thirties, with a still-crippled left hand and a noticeable limp, she made an astonishing claim: "Without hesitation I can say that losing my sense of smell has been more traumatic than adapting to the disabling effects of the stroke." Zierah described more than just the pain of pleasures forever lost. Like Michael Hutchence, she felt detached from reality, as if she were adrift in time and space. And her depression seemed to be deepening.

The novelist Vladimir Nabokov wasn't writing about depression when he confided in his memoir *Speak, Memory,* "I do not believe in time." But depression does teach one how tenuous is the sensory convergence that keeps a person believing, intuitively and without question, in a concrete and well-integrated version of reality. When that faith is eroded by a sensory breakdown, the rebuilding process can take years.

As my days passed without smell, I lost faith in both clock and calendar. My life was a puzzle that had been shoved off a table and onto the floor. It lay in pieces. I could reassemble it—it would just take time, whatever that was. Each piece was small, unrecognizable, and insignificant, but like the individual odors that make up a smell, each was programmed to contribute something to the whole, something essential and precious that I'd misplaced but would recover, the way you can unexpectedly stumble on a lost necklace or misplaced keys. You believed the keys were gone for good when in fact they were right there, within your grasp. The keys always show up. The pleasure of routine would come back the way feeling comes back after you've sat on your haunches too long and your feet have fallen asleep.

And yet . . . how do I measure the emotional cost when I realize I've tasted coffee for the last time? When I know I've taken my last stroll through the garden just to smell the fragrance of thyme that's released with every step and the roses that always entice first my nose and then my touch with their lurid perfume? When I realize I've smelled the last whiff of my grown-up daughter's drugstore-bought scent? (Never again will I suggest she tamp it down a bit because it makes my nose itch.)

What happens when I'm no longer held captive by smells powerful enough to take me far away in time and space? To my kindergarten classroom (oilcloth and crayons). To the state fair (cotton candy and horse barns). To the North Woods, where I finally got to know my true self by getting in tune with nose-tickling lichens and liverworts, ferns and fungi, and the tall white pines and the long shoreline of Lake Superior, with its pebbled beaches, sculpted driftwood, and the seaweed that always reminded me of a mermaid's silk tresses, sending up its smell of unknowable realms below as it swished against the sun-dappled stones at the water's edge.

What happens when physical intimacy is oddly arid because I can't even smell my husband?

How is one supposed to feel about a disorder whose only symptom, once the fake smells give up, is no sensation at all?

9

TASTING THE HOLIDAYS

ONE MORNING as the holidays loomed I awoke as usual in a vile stench and was (as usual) startled when the toothpaste set off a defensive dead-fish counterblast. My tongue tried to slither into the back of my mouth. Determined that Christmas would be business as usual, I made the annual holiday pilgrimage to the bakery across town that sells an anise-flavored Swedish rye bread and pastries to die for. The best part is lingering outside to admire the gingerbread village in the bakery window—all candy-cane roofs and gumdrop door-knobs—while catching hints of the olfactory pleasures waiting inside every time a customer enters or departs in a cloud of warm, sweetened air.

A tinkling bell announced my arrival; this time I hadn't even glanced at the window before crossing the threshold. Inside the crowded store, my nose reacted with dismay. There was no mistaking that the better the actual smell, the worse its surrogate. Other notoriously stinky places flooded my consciousness—the old Waldorf paper plant that ground newspapers into pulp and was finally shut down because of its stench; the oil refinery that had burned to the ground, incinerated by its own disgusting smell; and the Landmark brewery, which smelled exactly like burning toast soaked in stale beer until its odor had the audacity to invade the "better" neighborhoods and caused such a stink that the brewery finally had to install filters.

As I instinctively raised my hand to cover my nose, the sweet strains of "The Little Drummer Boy" gently nudged my thoughts away from the salacious apricot Danish, flaky kolacky, and caramel pecan rolls to the petits fours. How pure they looked, their smooth icing topped with dainty pink flower buds. On impulse, I added two dozen of the chaste little cakes to my order. Simplicity is the essence of a petit four's taste as well as its appearance. Butter, sugar, eggs.

My tongue still worked. Maybe the pure, sweet taste of a petit four could slip by my nose. I left the store, lifted the lid of the box, and pulled one out, then licked the top. Smooth and, yes, sweet. No question about it. I let it melt in my mouth while I tried to conjure up the taste. Butter, sugar, eggs.

But something was throwing it off. Confused by phantom smells, my tongue couldn't think straight either. I quickly swallowed the rancid lump and left the rest of the petits fours alone, unmolested, their virginity intact. Christmas Eve would bring another onslaught of aromas twisted beyond deciphering. Candles, wine, the roast tenderloin I'd most likely leave on my plate (a treat for Mel). Even the evergreens heaped in the center of the table would reek.

Sweetened foods are the most tolerable when you can't smell, probably because sugar is the least ambiguous of tastes and utterly without flavor. It is also the likeliest to come in the form of smooth and soothing, reliably unambiguous textures. Yogurt, ice cream, applesauce, certain fruits . . . petits fours.

To impress on her psychology students the relationship

between taste and smell, Rachel Herz had them taste jelly-beans with their eyes and their nostrils closed. Without being able to smell anything, the students found that the candies all tasted identical. No differentiating flavor. When the students unplugged their noses and popped the candies into their mouths—still no peeking—they were able to tell lemon from licorice, and bubble gum from grape.

Taste is said to be 90 percent smell—but only in humans. The secret to savoring, and why only humans do it, is flavor. *Taste*, then, is a misnomer. Some animals have lots of taste buds (and many different kinds), and some animals have none. Dogs can taste sugar but not salt. While humans can enjoy and even identify tens of thousands of *flavors*, we're able to detect only five *tastes*. Each taste has its own patch of tongue real estate—if the tongue were a residential lot, the bitter-taste taste buds would hang out in the backyard and the sour-taste ones in the shade on either side of the house. Salty and sweet tasters would be those more gregarious types clustered around the front gate, waving at complete strangers.

The taste buds depend on smell to make bitter, sweet, sour, salty, and umami worth much. (Umami is monosodium glutamate, or MSG, common in Asian food and often blamed for headaches by Americans. Some scientists now think that American brains are more likely reacting to its novelty rather than to anything innately allergenic or potentially harmful in the chemical itself.) Manufacturers of pet foods hype their flavorful doggy treats with the human consumer in mind. Obviously, no dog in its right mind is going to turn up its nose at a sweet or the remains of your dinner, although it's all the same to the dog if you had meat loaf or chicken tetrazzini. One olfactory expert put it this way: "When it comes to tracking the scent of a gazelle on the savannah, we can't compete with our hounds, but once we drag it back to the campfire we can sure season the hell out of it."

In humans those five basic tastes deliver the perfect gustatory package to complement the odorants rushing up the nose to the olfactory bulb. Psychologist Richard Stevenson called this associative learning. A couple of exposures to a sweet taste, and the nose quickly deems it sweet-smelling.

This also happens in reverse. When someone refers to the fishy taste of New England clam chowder, he's mostly reacting to its smell, a blend of cream and butter heavily seasoned with cooked onions and the smell of the sea. The tongue detects only the soup's sweet or salty sensations, as well as its warmth and the delightful feel in the mouth of slightly rubbery clams and soft, squishy potatoes. Neuroscientists now suspect that smell's tutorial role in tasting is facilitated by a synaptic interplay—a network of neurons speaking to one another. It's almost as if each of us has his or her own staff of expert chefs fiddling around in the brain, sipping, stirring, and spicing.

Chewing greatly enhances the flavor of food too, as some superintelligent caveman found out during the Pleistocene epoch. This would have been around fifteen thousand years ago. Mother Nature had done some plastic surgery to make the human mouth suitable for chewing tender (roasted) meats. Chewing made food more nutritious, easier to eat, and better-tasting even as it made human faces more refined—no more fanglike incisors and protruding jaws. Roast of rodent was just the beginning. Vegetables followed. These were roasted too, spiced with herbs, and blended into threshed, cleaned, and cooked rice, beans, and barleys, complex dishes inspired not by the taste buds but by the addition of the much more refined and discriminating sense of smell. The advent of farming added a huge array of new smells, from the enticing aroma of baking bread to boiled mush, butter, cheese, and alcohol.

It's not that other species lack the hardware—a dog's nose is enormous, and odors are just as accessible to canine brains as they are to humans'. It's a software problem that puts the pleasures of a fine Bordeaux out of Mel's reach. Imagine if you had to run your fancy new laptop on pre-Windows DOS. Or use a primitive word processor like TextEdit, which gives you a box to fill up with words and some rudimentary grammar tools, instead of the latest version of Microsoft Word, which corrects your spelling, keeps track of page numbers, red-flags your run-on sentences, and even lets you drop in footnotes.

Luckily for us, as neurobiologist Gordon Shepherd wrote in a 2006 article in *Nature,* "a notable share of this enlarged brain capacity is involved in flavor perception and behavioral responses to flavors." But like a complicated computer program, the human palate requires an investment in time and training to operate at its full capacity. In cooking too you have to learn the program. This process is both individual and cultural. With training, a person can learn to like just about any food, even raw fish.

A taste for sashimi comes so easily if you happen to be Japanese, though, that it's practically innate. Indeed, an ongoing debate in culinary and olfactory circles involves just this question: Are certain taste preferences hard-wired for certain peoples? While the answer seems to be no, cultural conditioning has a huge impact and may seem hard-wired. We know that vomit can smell good or bad to people depending on the culture and context—because, as a molecular matter, vomit is also Parmesan cheese, a food that is enjoyed by Western cultures but reviled in the Far East. Moreover, while a dog will happily lap up its own vomit, humans generally don't. The French like smelly cheeses, but many hygiene-obsessed Americans can't get past the mold. Fermented fish sauce and wintergreen are two other culturally dependent aromas. Asians love the former and hate the latter; with the Brits, it's vice versa.

Generally speaking, children are more open to new smells than adults are, and are thus more inclined to like the basic spices of several different cultures, which is why so many American kids can dine on pizza or tacos one night and lamb couscous or California rolls the next. In the span of a few decades we've evolved from a meat-and-potatoes (hold-the-onions) society to a multicultural one. But these positive feelings are ingrained early; even in the womb, the unborn child is learning to like what Mom likes. So it's not that kids are open-minded; they're just more easily trained because their brains are still developing. And yet, most children are timid about trying new things. All species will reject the unfamiliar outright the first time. With humans there are second chances. And third chances. Sometimes it takes a dozen exposures (in pleasant surroundings) to train the nose to change its mind.

Foods can have profound emotional connotations. We humans don't stray far from our cultural food groups unless the new dish contains a critical mass of the old familiar smells and tastes. No one is born with an exceptional palate, meaning an inherited sensitivity and openness to food. Even the most adventuresome eaters are trained.

But again, the American distaste for certain Asian delicacies such as fish eyeballs isn't set in cement; it's ordained by those coconspirators in the limbic system, memory and emotion, responding to novelty. What's new is threatening. It might be dangerous. Curl your lip, squinch up your nose, and return that eyeball to your plate.

Those facial expressions denoting disgust, by the way, probably *are* hard-wired — not the reaction to the food but the reflex itself. Chimpanzees communicate this way all the time. So do infants, even though the look is not directed at (or triggered by) their own dirty diapers. Feces are an acquired distaste. To what extent that distaste is developmental (inevitable), we can't say for sure.

New work in genetics isn't making it any easier to separate nurture and nature. Scientists have known only since 1991 where the smell genes hide, and it's possible evolution is responding pretty fast to cultural changes. If genes are this fluid, is anything really hard-wired? Most humans (and all Neanderthals) used to be lactose intolerant, but when people in cold climates started drinking milk (which was necessary, since their diets lacked adequate vitamin D), nature selected for a mutation in the DNA that essentially switched off lactose intolerance. Caucasians still tend to be lactose tolerant, and most Asians can't metabolize dairy products. (Could that be part of the reason for their revulsion to Parmesan cheese?) Maybe a cultural shift spurred by a nutritional need can affect the olfactory gene pool. Again, we don't know.

We do know that as the human nose became more neurally complex and as its relationship with memory and cognition grew more nuanced, it became a bit of a snob, perfectly capable of rejecting the wine if its bouquet wasn't just so. It reveled in the addition of spices that enhanced flavor. Indian, Mediterranean, Chinese, Japanese, and Hungarian cooking each has its own signature spice blend containing just a few key ingredients. In Hungary, the staple flavoring agents are paprika and lard;

in the Mediterranean, garlic, rosemary, lemon, and olives; in China, rice wine, soy, and ginger. Some scientists speculate that even though humans are losing smell genes over the long term, those that are still active represent a hot spot in the human genome and are evolving to respond to extremely specific nuances in various meats and seafood as well as beer and wine. Our species' recent willingness to cross cultural lines suggests that nature and nurture are, as usual, working together to make life more pleasurable and interesting.

Robbed of simple yet indispensable taste pleasures such as sushi, salsa, basil pesto, fresh walleye (fried in butter over a campfire), sole amandine, corn fritters, popcorn, and foie gras, I couldn't help wondering why there were people for whom these foods were meaningless and sometimes even loathsome. Is it the taste buds, the olfactory epithelium, or the central brain that insists that béarnaise sauce smells like stinky fish (or something equally off-putting that can't be put into words)?

A friend of mine noticed subtle oddities in his autistic son's behavior even in infancy, long before his later problems emerged. By the time Sammy was old enough to feed himself, issues surrounding food were front and center. Why wouldn't he eat anything? He lived on chicken tenders and chocolate milk. Why such children find many foods unpleasant may stem from the same synaptic misfiring in the limbic system that interferes with their ability to learn from positive experiences.

Autism makes it difficult for the brain to erase old food associations and replace them with new ones. When an autistic child tells you that the delicious cheese omelet you've made just for her tastes like glue (or worse), she's not kidding. It does . . . to her. And the glue factor is as real as the stench of phantosmia is to someone suffering from a damaged olfactory system.

Autistics do overcome some of their aversions some of the time, but the learning process can be as frustrating and illogical-seeming (why does she refuse eggs but not spinach?) for the child as it is for the parent. Social problems are so much more painful than any difficulties over what's for dinner that underlying parallels are often overlooked.

The day before Christmas I lifted my head from the pillow and sniffed. Nothing. I tried to summon a particularly noxious scent: burning rubber. *Come and get me. Whattya, chicken?* Still nothing.

Rotting fish ... roadkill ... skunk.

No, no, and no. I couldn't remember what they smelled like beyond the verbal descriptions I'd crafted for anyone interested in feeling my pain when the stench had been constantly present. My head was clear. Phantosmia had left without a trace. Alex and Caroline were already setting the dining room table for our dinner guests when I came downstairs. "It looks so pretty," I said, trying to ignore the demons that were whizzing about the room like those flying monkeys in *The Wizard of Oz,* telling me in their leering, nonverbal (but all the more sinister for that) way that I'd entered a whole new kind of hell. So this was anosmia.

I made my to-do list while sipping the tasteless coffee that explained the more pronounced than usual tremor in my hand. Nerves. Caffeine. Whatever. This afternoon I would wrap presents, mend a tear in the sofa, polish the silver until it gleamed. I would, of course, leave the cooking to my capable husband, all but the chocolate roll, which was a family tradition and not easy to make on the first try. The roll wasn't much to look at but it would taste good ... I hoped. I'd never know.

What should I wear? My clothes hung on my skinny body.

I didn't want anyone to notice, to know, to talk about it. I could control this. If I didn't seem bothered, other people wouldn't find it horrible or freakish either. They wouldn't find it much of anything. A flu bug. A hangnail. A perfect circle of deceit. Put on a smile, my mother always said on her way out the door to the hairdresser's after some small setback, and your mood will follow.

My daughters sensed their opportunity—Mom's willing to let us have a go at her. Alex rummaged in her closet for a stylish blouse and a pair of slacks that wouldn't have fit me just a week ago.

"I seem to have lost a few pounds," I said. "How fabulous is that?"

Caroline circled with the eyeliner and mascara. I removed my glasses and gingerly placed a contact lens on each eyeball, a trick I've never quite gotten the hang of. The girls discussed blush for fully ten minutes, using sign language to communicate decisions regarding wrinkles and age spots (to cover or not to cover?). Finally I was ready for the finishing touch. I still looked wan and tired. My mouth was their last shot. They fished around in their makeup bags. I told them that I would bow to their judgment. Lip-gloss, I said, is fine with me.

Alex pulled out a frosted coral shade and applied it to my mouth. Caroline shook her head and suggested dark red.

No improvement. "How about mauve?" she said.

"You'll see," I said, shaking my head. And they did see. My small pink mouth became a mauve slit.

"Let me show you something," I said. After rubbing my lips with a tissue, I slowly removed the top from my own lipstick. I twisted the tube to raise the bright orange missile out of its silo. It still had the perfect body of a brand-new lipstick, the fingernail-shaped tip, the flat, smooth place that meets the lips head-on. If you were under forty, the slightest hint of color above the lip line suggested a secret wish to have lips as full as Angelina Jolie's. Over forty and it meant you needed bifocals. I was perilously close to the age when lipstick that strays beyond its appropriate boundaries settles in the vertical creases made by decades of pursing one's lips. A deep frown line in my forehead was my skin's only obvious stress fracture, but the lines along my upper lip in particular were quite visible if you were really looking for them—another reason to save the broad, indefinite smear for later.

Suddenly I understood why old ladies love lipstick. What do they care if they overindulge? The more lipstick used, the stronger and more long-lasting the nostalgic perfume that goes up the nose with every inhale. Lipstick is better than Botox for making a woman feel young and sexy. Cheaper too. I now realized why each daughter's attempts to color my lips had failed to induce in me the slightest pleasure. It wasn't so much the visual result that was a disappointment but the olfactory absence. It was my lipstick's smell, as strident as its color and just as eager to please, that had gotten to be habit-forming. That fake fruitiness filling my brain was the smell of winning people over, getting my way. It made me confident because as long as that wonderful fragrance hung around (always too briefly), I could be someone else, someone infinitely more desirable and competent than I really was. When the smell faded, I was my plain old self again for all the world to see—until I could get to my purse for another stiff shot of the buoying elixir.

My current lipstick was by Revlon and called Peach Sunset. Appropriate. I applied it in one fluid stroke across the upper lip and then the lower. Then a gentle dab at the twin pyramids just below the nostrils, and finally the ritual pressing of the lips together around a tissue and a quick glance at the soft bright replica of my mouth before my fingers flicked the tissue into the wastebasket. Without the tangy aroma to trigger the ego boost I attached to the lipstick, I could feel myself shrink into drab, tongue-tied obscurity.

"You know, you're right, Mom," Alex said. "Orange suits you. You look great."

Alex buttoned my blouse—her blouse—and I slipped on my (her) pointy-toed high heels. The dressing room was littered with hairbrushes and undergarments as well as fragrant lotions, lipsticks, hair gels, and perfumes, and yet to me it felt as barren as a white-tiled tomb.

At the stroke of seven, the relatives burst through the front door in a pack, red-cheeked and beaming. Four families (including ours), twenty-some children, my mom in her wheelchair. Everyone exchanged hearty hugs in the front hall. Ribbons streamed from shiny packages. Steam rose from my sister-in-law's special pizza rolls; everyone was thrilled by their sharp pepperoni smell, by the smells wafting in from the kitchen (tenderloin in the oven, garlic mashed potatoes on the stove), and by the winter coziness of the crackling fire and the fragrance of the Christmas tree and the pine boughs.

I collected coats and parkas and brought them upstairs, where they'd spend the evening in a congenial heap on top of our queen-size bed. Suddenly exhausted, I flopped down beside them. How I wished I could stay right here, where it was quiet. My family's cheer made my own feelings of estrangement all the more acute. *Where've you been?* they would ask as one when I went downstairs. *How* are *you?*

They would laugh if they knew. They would roll their eyes and say, *What would she do if something really serious happened?*

I lay on the bed and thought about my mother. Even now, entering her daughter's home in a wheelchair, frail as a sparrow and facing an evening that would slowly erode the forced merriment she'd trained herself to exude over a lifetime of being the perfect hostess, she could arouse my envy and awe. It was she who taught me that "putting on a smile" meant wearing lipstick. Tonight she had on a smart red wool suit that matched the lipstick on her wrinkled mouth. She'd been to the hairdresser. Her nails were also painted red. Even though she was weary and lightheaded — always — and hated going to a party in a wheelchair and wondered if there would be anything bland that she could eat, I didn't need a nose to know that she'd taken the trouble to put on perfume; out of habit, yes, but also to honor the holiday. There was far too much noise; too many children, dogs, presents, hugs. But it was Christmas, after all.

Mom was dimly aware that I had been "ill." She wondered why I hadn't gone to bed to get over what was ailing me. My sister, Judy, explained to Mom that anosmia wasn't the kind of illness that sends a person to bed. I was having trouble smelling. "Is her marriage all right?" she whispered to Jude. Mom's had not been all right. Something smelled funny to her about happy marriages, she always said, or rather hinted. Most of Mom's honest beliefs were not articulated, even to herself. But she had ways of getting a point across. The collapse of a marriage made perfect sense; the failure of a nose she could not fathom.

10

IDIOT SAVANT

WHEN I'M FRIGHTENED I'm at my most acute. Twisted, but razor-sharp. I'm like smell in that way, the sense that smell expert Trygg Engen so aptly described as "an idiot savant." Smell's keen awareness, its acuity, became more and more apparent as I tried to get used to living without it. Before this, I'd never questioned the almost universal assumption that smell is the least important sense. How absurd such judgments are. You can't rank the importance of an essential part of what makes you not just a human being but a unique individual, the sum total of all that you've experienced through your senses. Besides, I was beginning to realize that in its own quiet and mysterious manner, smell actually knows more about many things than the eyes and ears do. And that in some ways the human sense of smell is more perceptive than that of animals who live and die by their noses. What humans live and die by is just more complicated. Since we don't live in the moment, mere survival doesn't suffice to make life worth living. Smelling has evolved in our species to enhance the experience; it helps us tune out the ticking of the clock by adding nuance, texture, and an emotional component that is so exquisitely subtle we're not even aware of it.

The day after Christmas, Caroline asked if I'd like some help cleaning the house. I declined the offer. She had better things to do with her vacation, I said. She guessed correctly that I wanted to stretch out the household chores so they'd last until it was time to make a semblance of dinner. Leftovers. No actual cooking (that is, the smelling and tasting and testing for the right mix of ingredients) required. As she was leaving to meet her friends, Caroline mentioned to me a scene in John Steinbeck's *East of Eden* that had helped her get through some rough patches in her own young life. In the novel, a wise neighbor comes to the rescue of a jilted husband, not through lengthy talk sessions but by repeating the same gentle instruction over and over: "Go through the motions, Adam."

Caroline told me, "You'll get used to this, Mom." Then she repeated the line from the book. "'Go through the motions.' Just put one foot in front of the other." Soon every day would be just another day without smell. That was something to live for.

I collected the broom and sponge mop, my bucket and rags, the spray cans of Lemon Pledge and mildew remover, the toilet cleaner and the Comet, and I went to work on the house. I decided to take a yellow pill, not to make the work easier but to quiet the unceasing clamor of loss. How I'd loved cleaning my house before . . . this. Physical activity of any kind had a stimulating effect. Now, it seemed, the stimulant was missing. Had it been the smell of cleaning products? Just because I hadn't been aware of them — the citrus in the furniture polish, the chalk in the Comet cleanser, the chlorine in the mildew remover, and even the putrid old-shower-curtain smell of mildew itself — didn't mean they hadn't been there, urging me on.

I couldn't stop thinking about them and all the other smells of housework. When my damp rag passed over the spine of some ancient volume in my great-grandfather's library, I remembered how I used to slow down instinctively when cleaning the books, lighten my touch. Dust rose from the cracked leather binding and I saw that I'd disturbed the thin layer that was like a fine sand on the top of the book's unopened pages. A funnel cloud of particles caught in a shaft of sunlight lit up like thousands of bright golden bits of fuzz rising on the thermal that I'd created just by sniffing. And along with the dust came whatever it was that bound together the smells of old glue and leather and the kind of paper that stands up to age without turning yellow except around the edges, where just enough oxygen can sneak in and weaken certain chemical bonds and thus the paper itself.

Though before I lost my sense of smell, such moments were not rare, I'd never deconstructed them—those times when I'd put down my rag, pull a book from the shelf, and open it. I'd stumble as if in a trance, my nose buried in the book, my head full of curiosity and tenderness for this old but still intact treasure, to one of the chairs that had also belonged to my great-grandfather; I'd slowly pivot in front of it (still not lifting my eyes) and back up so my calves were touching the seat cushion, and then I'd flop down like a cushion myself, pull my feet up, tuck them under me, and scoop Mel up into my lap.

I tried to remember the strong smell of the carpet, hoping this would set off a ripple effect. From dogs to dust and dust to leather. Can those olfactory neuron maps stored in the hippocampus (whose shape resembles, of all whimsical things, a sea horse) be brought into consciousness without the chemical knock on the door that alerts the brain to smells in the upper nasal cavity, smells that are let in only after they've spoken the secret genetic password? And how about smells that never make it to the level of conscious thought, those nameless odors we don't even know are there?

Staring at the worn carpet failed to bring back its peculiar aroma. I seemed to have forgotten it, maybe because it was not altogether pleasant. Good smells are said to stick around longer than bad ones. I tried another: the smell of the worn tapestry fabric on my great-grandfather's wing chair. I sank into it, closed my eyes, and tried to summon its odor, this time by sniffing. I thought about how the chair had sat beside the fireplace for as long as anyone could remember. The comfiest chair deserved the warmest place. It had absorbed the odors of slow-burning oak and the silver birch logs that went up like parchment; each tree as it burst into flames gave off its own distinctive whiff of cell structures being violently rearranged. The trees left their imprint on the chair; the smells of their incineration seeped into the pores of its upholstery, and into the feathers inside the cushion, and into the wadded-up cotton stuffing added later and now gray with age, and into the springs and the thin fir slats that held the fabric to the heavy-timbered frame that sat on four claw feet. Still shiny from layers of varnish, the feet left hollows in the carpet when the chair was moved for the semiannual deep cleaning of the dust, accumulated buttons, pennies, and pine needles that hid beneath it, out of reach of the vacuum nozzle.

Yes, the wing chair was a sponge for smells. My sensitive nose always knew who'd sat in it the last time we'd had company based on hints left by perfume or shampoo, or just the odor that comes off a person's skin, as unique as a fingerprint. Human odor is more pungent in times of stress or excitement.

The scent was always more readily detected if it had been warm in the room and there'd been an argument.

It didn't make sense that the organ that made such subtle distinctions and attached even more subtle emotional messages and responses should be regarded as inferior to the same organ of a species that was led by the nose in direct and obvious ways. A wine is judged by its complexity. Why not a sense? Dogs are glibly pronounced superior in all aspects of smelling. True, a canine's world is saturated in smells, and a dog lifts its nose to catch scents from miles away. A dog is able to distinguish the smell of its owner from that of a stranger and to tell other dogs apart. We're so impressed by such seemingly uncanny talents that we all but ignore our own smell acuity. Human olfaction isn't inferior to dogs' but different, just as humans differ in having bigger and more complicated brains. As the human brain evolved, its older parts received upgrades, abilities made possible by the intricate wiring of the newest features.

The retronasal passage, located behind the mouth, allows all the intermingled aromas of one's surroundings, including some released through chewing, to slip up to the brain by the back way. To neurobiologist Gordon Shepherd, this suggests that the central brain contributes more to smell function than the peripheral parts do, which in turn explains why, in his view, humans actually have an excellent sense of smell in spite of our small noses and declining number of smell genes. The primary sense seems to have insinuated itself ever more securely into the psyche via those complex neuronal structures of the neocortex. Even rats have been shown to smell just fine after 80 percent of the (peripheral) olfactory system is removed.

Shepherd thinks that our active smell genes are becoming

more focused on the specialized olfactory needs of a highly evolved species. It's not the overall gene count that matters; it's what each gene does. Smell scientist Stuart Firestein discovered that one human olfactory gene performs the work of three or four mouse olfactory genes.

That humans smell quite well is more than just a fun fact to toss around over lunch in the lab cafeteria. Take a recent contest dreamed up by psychologists at the University of California at Berkeley: canine smellers and college undergraduates both tracked a trail of chocolate on all fours. True, the dogs outperformed the kids by a factor of four. But the students surprised the scientists by how few mistakes they made. With their noses pressed to the ground, as dogs' usually are, the students were able to smell the relatively big, heavy odor molecules that escape detection when one is standing upright.

Moreover, what the human nose lacks in raw olfactory firepower, it more than makes up for, Shepherd argued, in finesse. This is why humans may have a better sense of smell than animals that have many times keener noses — better for what humans enjoy doing, which is not wading into mosquito-infested marshes to grab a dead mallard but rather shooting mallards and then handing them off to a good cook to create a meal that would be wasted on a golden retriever.

That women score higher than men on tests of smell acuity seems to contradict the finding that all people are equal when it comes to smell. But the nurture side of the nature-nurture polarity is expanding its territory. Biologists now count prenatal influences as nurture, not nature. This is important. They want to be absolutely precise when determining if a given biological difference is gender-based. Since genes can be altered in the womb by elements of the outside world, even a genetic difference isn't always nature-driven—that is, determined more by gender than other factors. Boys are stronger than girls. This is a biological fact. Are boys smarter than girls? This is a minefield. But the same brain research that is pointing us toward a negative answer (no, they're not) is behind current thinking on smell: girls are better smellers. Not born better, but primed to want to be better at it and thus more inclined to learn. I'm not talking about conscious learning, as in a classroom, but conditioning that comes from experience and gender roles.

Why was I so aware of the smells now missing from my house, many of which I'd noticed only vaguely when I could smell? Would my husband have had the same heightened awareness of smell's absence if this had happened to him? No. I doubted that he would have gone into as deep a funk as I had, nor would he have been so keen to deconstruct smell (as if that would bring it back). He would have acted like a man, sucked it up, and moved on.

This fact struck me as the key to my original question: are women better smellers than men? Only to the extent that they have evolved to perform gender-specific functions that seem to be part of their DNA because of adaptation. Unhelpful genetic traits are extinguished by the death of the individual burdened with them. Helpful traits live on. In the latter category in humans is a nose that is highly impressionable and easily trained. Then life experience takes over.

Such thoughts occupied me as I cleaned. I was keeping a tight rein on thinking. Keep it simple. Logical. Focused. I am a classic random personality, eager to chase every capricious notion that enters my brain. This habit of thought (and behavior) is why my house is usually a mess. *We'll have none of that,* my inner Nurse Ratched scolded as my thoughts wandered. She was the polar opposite of smell. Cold, practical, and as sensitive as a post.

I'd lived a third of my life in this house. I considered it part of my DNA. Or maybe I should say my family's DNA, seeing as how I was the fourth generation to live here and the house hadn't changed much since it was built. The wing chair smelled of me and Mel and a thousand meals enjoyed before a thousand fires and whatever else I'd contributed to the second skin it had acquired since my husband and I moved in. We'd had it reupholstered and the springs fixed fifteen years ago. The heavy fabric that replaced my grandmother's pastel chintz gave the room an earthier, more masculine feel that also disguised the fact that we don't have live-in maids as she did. All the chairs were going to see more wear living with us than they had in three generations with my ancestors. The wing chair would acquire the slightly oily aroma of cat hair hiding under the dilapidated cushion, which was now as limp as a flat tire. It also held the aroma of Sweetie, the black standard poodle we'd bought as a puppy for the girls; thirteen years later, she reeked of cancer and could barely get up into the chair she used to leap into with the grace of a fawn and loved to lounge in with her long black paws hanging over the edge of the cushion like a pair of slender silk tassels, her long elegant nose resting daintily on the soiled arm.

All these moments would soon be imprisoned in their fabric cage, their smells locked away in the limbic system. I closed my eyes, determined to resurrect others before they too disappeared without my saying a proper goodbye. The chair captured and held them, an eyewitness, though its clues could not be deciphered through vision. Only through smell. The curtains and the drab pea-green carpet grabbed the smells too: compost and the crisp fall leaves and bits of grass that stowed away in dog and cat fur and came into the house through the rubber flap of the small opening in the side door that I'd cut out years ago with a Sawzall to let pets go in and out of our fenced garden at will; the countless times wine had been spilled, or coffee, or milk; the smoldering embers the fire spat up and over the screen, leaving a burned-carpet smell and a black spot; and the dust from the books in the bookshelf.

My house with all its marvelous smells was my refuge, my joy and comfort, the only place that could persuade time to slow down, back off, and sometimes, at very special moments, like when I was cleaning and became distracted by a book, even stop altogether. I could revel in the house's sensuality, not just the smells but the images they summoned, the people and moments when we had all been so much younger.

I opened my eyes. The smells, such as they were, vanished. What would my home be like now? Not just this chair, but my garden? My kitchen? This room stripped of its layers of fragrance was a barren cell. It would no longer delight my eye, its physical beauty having been formed by the smell of it. I might as well live in one of those new condos that Cam suggested we take a look at now that the kids were gone, with their white Sheetrock walls and plastic carpeting and sanitized air. No dust, no dirt, no crumbs allowed. No pets or children. No memories.

What did this room smell like now that the bad odor was gone? How do you describe what isn't there? Especially something that is off-limits even to memory, trapped forever in some shuttered chamber deep in the brain? The brain can store pictures and events, and melodies, and even smells, especially smells—but like Sleeping Beauty, who can only be awakened by a certain kiss from a special prince, smell and all its attendant emotional connections lie dormant until triggered by smell itself. This requires a working nose. Without my sense of smell, my great-grandfather's wing chair would gradually become a stranger. Its visual presence would mock the richness of what it had once been to me. It would be like the photograph of a loved one who has died. Wouldn't you rather, until the pain has dulled a bit, put the picture away?

11

SENSELESS EATING

B ECKY PHILLIPS SAID it took months for her to stop dreading meals: the shock of recognition "with each and every bite," as she put it, that steak was no longer steak but something profoundly and repugnantly altered. In other words, while anosmia might come as a relief from phantosmia's twenty-four/seven assault of putrid smells, it would not restore my pleasure in eating.

When you can't smell, the simpler and more familiar the food, the more traumatic its transformation. Buttered toast, the ultimate comfort food, is no longer familiar. You've been conditioned through long years of exposure to expect toast's symphony of delightful sensations, beginning with the sweet aroma, then the liquid saltiness on the tongue, and finally the crisp yet chewy feel between the teeth. There is no way to describe the revulsion that attaches itself to the poor excuse for the food that every fiber of your being is begging for. To compare the taste of scentless toast to cardboard is to admit defeat.

I'd been smell-free for a month when Dr. Cushing took me off the amitriptyline. I assumed the stench would return. My hus-

band assumed the opposite. To him, the end of phantosmia sig-
naled recovery. He was convinced that my sense of smell was
on the mend—until the night I tried to cook dinner for a
dozen guests. Lasagna seemed like a no-brainer. How could I
screw that up?

Cam never said the flavors were off. He didn't have to. The
evidence was on all those confused faces around the dinner table, and on those half-full plates, and in the second pan that I
always had ready for second and third helpings. It's hard to
cook well when you can't smell, almost as hard as it is to eat
when you have no appetite. The extra pan went back in the
fridge untouched.

No one asked for seconds, even after I'd helped myself to
thirds.

That was the weirdest part. I couldn't stop eating.

"I know, it doesn't make sense, does it?" Becky Phillips said
after telling me she had put herself on a diet. "You'd think anosmia would be a great diet plan. But now when I'm eating I lose
track of what I'm doing. I don't ever feel full. It's as if I don't
know that I *am* full. And of course I never feel satisfied. So I
just keep on eating, expecting to feel like I've had enough, but
that signal never comes. I used to think that feeling full came
from having a full stomach. Like 'I'm so full I'm going to burst.'
But apparently it's something else."

By late February I'd gained back all the weight I'd lost, and
then some. This should have been a sign of recovery, if not of
my sense of smell then at least of my joie de vivre. It wasn't. I
had not adjusted to anosmia. Acceptance of anosmia can't be
taught in a self-help book. I'd tried that too. Lots of books deal
with how to cope with loss, but my loss felt ludicrous compared
with the ones in those books. My rational self still scoffed at my

emotional self for thinking that being unable to smell was in any way comparable to losing a loved one.

I was traumatized and alone with my imponderable pain. I was still trying to ration those little yellow pills that seemed essential to getting me through the day.

Dr. Cushing hadn't prepared me for any of this—except for the weight gain. Twenty-five pounds is about average, he'd said. For the first time ever, I felt real empathy for overweight people. They'd always baffled me before. I'd thought, *If you don't want to be fat, stop eating.* It's not that simple. Overweight people complain that they struggle with demons. They crave certain foods. They can't stay away from sweets. They eat compulsively and can't help it. They try too hard to lose weight, which leads to yo-yo dieting, which leads to hormone imbalances, which create more of those addictive feelings about certain (usually sweet) foods.

It's those metabolic disturbances scientists want to know more about. Why do people feel compelled to keep eating even though they're full? Why do they compare the urge to eat to an addiction? What triggers appetite and satiety? These issues occupy the minds of biologists testing satiety responses in lab rats, and they also concern the physicians treating a growing number of patients whose problems, ranging from diabetes to sterility, have an underlying connection to hormones and appetite. These professionals have all reached the conclusion that the sense of smell plays a key role in regulating food intake. The primary sense doesn't just prompt the salivary glands to prepare the digestive system to make fuel from food. It also announces when the tank is full.

During his tenure as chief of the Food and Drug Administration, David A. Kessler, a pediatrician, accused the tobacco industry of studying certain human psychological vulnerabilities and deliberately designing products that made cigarettes irresistible to addictive personality types. Kessler thinks food companies and restaurant franchises do the same thing, creating what he calls hyperpalatable foods rich in a carefully balanced blend of sugar, fat, and salt—ingredients that just happen to be linked to the current epidemic of diabetes and obesity, as well as to cancer and heart disease. Texture and taste are paramount in such foods—think chocolate chip cookies—but certain microwavable instant meals also waft mouthwatering aromas into the kitchen. Kessler thinks processed-food manufacturers, probably without understanding the neuroscience behind their products, intentionally manipulate the brain's ability to create mental images of desirable things and then get stuck on them, especially in times of stress. To combat the nation's weight problem, he promotes a sweeping "perceptual shift." Nutritional education, he believes, has the power to turn the public away from unhealthy foods by literally turning their stomachs at the thought of them. The same phenomenon causes a once-avid steak eater who switched to vegetarianism after a heart attack to find the mere mention of a well-marbled rib eye repulsive. It happens with smoking too; after all, you don't have to be a neuroscientist to feel turned off by something you know could kill you.

Some biologists think Americans' lousy eating habits wreak havoc not only on the body but also on the olfactory system —specifically, that sense's role in triggering hormones that regulate appetite and metabolism.

Most Americans don't cook anymore. Not really. No olfactory signal of "doneness" emanates from the microwave oven as an instant meal is being heated. The normal sequence of hunger and satiety is interrupted when anticipation triggered by smell is removed.

No wonder so few families take the time to sit down at the dinner table and savor a meal. They haven't received those alluring multisensory prompts. We're like a nation of anosmics when it comes to family dinner rituals. Meals have been stripped of the essential anticipation: the gathering of appetites around the table as the food is served; the traditional sniffs over the steaming stew pot; the soup spoon sliding beneath the surface; a pause to savor the smell a final time before the mouth opens around the spoon and the soup is allowed to spill over the tongue. The irony is that even as we've developed an ever-higher food IQ as a society (we have access to and have developed a taste for many more foods than earlier generations), unless we're having guests for dinner or dining out, we gulp, nosh, and shlurp without much pleasure. Enjoyment and nutrition are both diminished, and obesity statistics climb.

Hormones tell us not only when to start eating but also when to stop. The limbic system includes the hypothalamus, an almond-shaped organ located below the thalamus and just above the pituitary gland; it regulates the nervous and endocrine systems. The hypothalamus is a busy place. It synthesizes neurohormones that stimulate or inhibit the secretion of pituitary hormones, which in turn control body temperature, thirst, fatigue, anger, and certain sleep cycles, as well as hunger. In people with working noses, the sensing neurons in the olfactory receptors and taste buds stimulate the hypothalamus to welcome an order of fries as well as tire of them, even though a few remain on the plate.

Adaptation (also called habituation) is why you can't smell the garlic on your own breath. It's what makes mice, and probably people, stop smelling an odor after a few minutes. But how do these mysterious mechanisms—satiety, adaptation—actually work? Investigators at Johns Hopkins University think a protein molecule in the receptor neuron causes a particular channel in the neuron's membrane to open and close, helping the brain discriminate among different incoming odors. An odorant triggers a series of molecular "gates" on the cell surface to admit certain charged ions. Differences in charge between the cell's interior and exterior allow electrical signals to travel the axon highway from the tissues lining the nose to every corner of the brain. Researchers speculate that the brain uses this same mechanism to aid in balance, fertility, and digestion.

About thirty minutes after ghrelin, the so-called hunger hormone, turns on the salivary glands, leptin and obestatin kick in to curb appetite. Destruction of one part of the hypothalamus can cause massive overeating, while destruction of another region causes the opposite to occur: appetite is erased. Recent research suggests that satiety occurs in two stages: first, the palatability of the food on your plate is selectively reduced (probably starting with the broccoli), and then food odors fade. Smell researcher Alan Hirsch tried to test this theory a few years back. His informal studies showed that sweetening food shortened the amount of time between hunger and satiety. He went so far as to market a sweetening agent, which he called Sprinkle Thin.

Venture capital is pouring into start-ups that are trying to come up with sure-fire weight-loss solutions that harness the body's own systems. Most research involves manipulating electrical currents that go to the brain (stimulating the satiety response in the hypothalamus) or the gut.

All too late for me. As I mindlessly munched on sugar cookies and taco chips, I felt like a robot programmed to feed itself indefinitely, experiencing neither pleasure nor pain. The robot would go on eating until its batteries died, unsatisfied to the end.

My husband insisted that I was overthinking my weight gain. It was hardly surprising that I couldn't stop eating: I was depressed. He always ate too much when he was depressed. Everything would return to normal once I got used to this.

Maybe so, but regardless of whether the hunger was psychological or physiological, it felt like a leaky faucet that would continue to drip until someone replaced the washer.

12

CULINARY ART

IN FEBRUARY, A FOODIE friend of mine named Peter told me that Grant Achatz, cofounder and head chef at Alinea, the Chicago restaurant considered by many experts to be one of the best in America, had been diagnosed with tongue cancer. Peter used to live in Chicago and knew someone on the staff who'd been keeping him up to date on the boss's health. Peter, Cam, and I had dined at Alinea together just six months before I lost my sense of smell. He hadn't called to bum me out, Peter said, but to invite us for dinner at his house the following weekend. Did I want that potato, butter, cream, cauliflower, and truffle thing we'd had at Alinea that Peter was dying to make, or would I prefer something less caloric?

I thanked Peter for asking and reminded him of my own situation; I said I could go either way on the dinner, and whatever he wanted to cook was fine. "Oh, sorry, of course," he replied. "I'll do the truffle thing. You can tell me if you like the texture."

Then it was back to Grant Achatz. "How unfortunate he didn't lose his sense of smell instead," Peter said, "like you did."

Better to lose smell than taste? Are you crazy?

Finally I said, referring to the cancer, "I sure wouldn't trade places with him."

I was lying. And not just because I'd come to value smell (in its absence) more than taste (what was taste without smell?), but because, for all the anxiety I'd had over the C-word just a few months ago, I was now battling something for which there was no treatment and that no one I knew, including Peter, had ever heard of.

In the latter half of the nineteenth century, a Frenchman named Auguste Escoffier was pushing the culinary envelope. He shifted the focus of the culinary arts from sight to smell.

The peasants of the time ate relatively well. They had no choice but to boil their tough meats (and to extend them) in a hearty and nutritious broth flavored with bone marrow, fat, and fresh vegetables. But eye candy—stunning, sweet-tasting but odorless confections served cold to preserve their beauty—was the culinary ideal in the royal court at Versailles. Antonin Carême, who cooked for Napoleon, was a superb visual artist. The only taste of much importance in his extravagantly ornate dishes was the sweetness extracted from rare and exotic fruits.

Escoffier radically altered the recipe for dining pleasure. Aromatic soups and stocks inspired his dishes; slabs of bone scented with herbs and spices and vegetables were simmered for hours, even days. Reduction with a dash of red wine turned simple stocks into enticingly fragrant sauces, which were served piping hot.

Born in Provence in 1846, Escoffier gave up his boyhood dream of becoming a painter to focus on a career that was more reliably remunerative but no less artistically challenging. Though he never attended a university, he understood that the enjoyment of food was part training and part biology. Instinct (or perhaps the word is *genius*) told him that his beef bourgui-gnon would taste even better if he engaged the diners' other senses as well. To turn novice eaters into discriminating gour-mets, Escoffier tapped into hearing and sight to enhance the pleasurable experience of his food. What he intuited was that the brain handicapped unpleasant memories. Clever evolution. Any mild displeasure at the dinner table would be selectively erased from the record if the *overall* impression was delight. Violinists strolled among tables adorned with gorgeous ar-rangements of fruits, flowers, and feathers. Not only was each dish perfect; so was each plate. Diners took unconscious mental note of the fine crystal glasses, the sterling silver candelabra, the crisp linens, and the smiling company.

Jump ahead to twenty-first-century cuisine: Chicago's Alinea. Even before the restaurant is entered, Alinea challenges every hard-wired expectation of going out to dinner. Located in a hip but by no means opulent neighborhood, it lacks outdoor signage (unless you count the tiny type on the valet-parking stand). Sliding glass doors softly swish open to welcome guests. The walls are gray, the restrooms black. Gray carpeting dulls the sound of servers' feet, a good thing, since the wait staff nearly outnumber the clientele and are not really servers so much as performers. A different one appears with each course, first to deconstruct the dish and then to advise on how it should be smelled and (with the help of custom-designed tools for which the word *utensils* does not suffice) tasted. The sommelier is busy too, flitting from table to table like a honeybee from flower to flower in a fragrant garden, presenting the wine spe-cially chosen for the course. The open-to-view kitchen is meant to be admired not for its hustle and bustle but for the Swiss-watch precision of the sizable kitchen staff, who work noise-

lessly at long stainless steel counters that give the room the look
of a surgical unit. Even during the mealtime rush, one employee
devotes all of his time to pushing a silent carpet sweeper up and
down between the prep tables, picking up any lint or lettuce
leaves that have fluttered to the floor. The impression of obsessive cleanliness belies what the place is about. Cooking is messy.
On the walls of the kitchen hang sketches created by the chef
to help his staff visualize—imprint on their souls, really—his
menu, which changes constantly. Ask for the restaurant's signature dish and you will be told there is none.

Achatz goes even further than Escoffier. To focus diners' attention on smell, the chef created a space where the simple dé-
cor is part of the game; it's as cleansing to the brain as the dish
called (with Alinea's usual understatement) Pear is cleansing to
the palate. Pear consists of pungent eucalyptus leaves and ber-
ries served in a small white bowl. A sliver of the glistening pale
gold fruit drenched in olive oil and black pepper is placed on a
spoon inside the bowl. This is what you slide onto your tongue
as the eucalyptus fumes waft up your nostrils. Long sprigs of
fresh rosemary entertain you with their scent during the brief
intervals when the servers are absent. Then it's on to Pork Belly
(with iceberg lettuce, cucumber, and a Thai distillation); or Bacon (a single slice suspended from a trapeze and eaten with
butterscotch, apple, and thyme); or Wagyu Beef with Achatz's
interpretation of A1 steak sauce (the spices are arrayed on the
plate beside the shimmering red cube along with Potato and
Chips). It is a relief to know that the chef—or Chef, as the
servers refer to him—can laugh at himself.

In Grant Achatz's olfactory cuisine, a concept invented in
Spain, smell becomes at times almost painfully delineated.
Restaurant-goers generally like to dine surrounded by the din
of clattering saucepans, laughter, and conversation—not in the

hush of a church. But Alinea's patrons bring their appetites there precisely because it is the temple of cutting-edge multi-sensory cuisine, the most pure and rarefied of dining experiences, the ne plus ultra of eating.

When we arrived at Peter's for dinner the following weekend, I saw a copy of Achatz's cookbook on the coffee table. Flipping through the glossy pages, I could only gape at the exquisite compositions and try to imagine their divine scent. In *Alinea-Mosaic,* itself a work of art, the chef reveals his secrets for creating such unforgettable experiences — there is no other word — as Bay Leaf Bubbles; Pillow of Nutmeg Air; Yolk Drops with Asparagus, Meyer Lemon, and Black Pepper; Licorice Marinade; Chicken Skin with Black Truffle, Thyme, and Corn; and Chocolate Warmed to Ninety-four Degrees. Kuroge Wagyu is a blend of cucumber, honeydew, and lime sugar, a nougat of olfactory bliss.

Another call from Peter a few weeks later brought alarming news. Achatz's doctor had said the tumor must come out — the tumor and the tongue with it. Instead, the chef chose a draconian regimen of chemotherapy and radiation that would destroy his taste buds, leaving only his sense of smell to guide him in the kitchen. During the months of his treatment he relied on his sous-chefs to taste and then communicate what they'd tasted to their boss. Using his sense of smell along with the sous-chefs' gustatory information, the chef was able to produce formulations — *recipes* is far too prosaic a term — for wondrously delicious and original dishes, proof positive that flavor is essentially derived from combinations of smells, with taste playing an important, but far less complex, supporting role.

Achatz's staff marveled at the chef's stoicism and discipline and were amazed by the consistently sublime dishes he created but could not taste. Achatz ignored the pain and his doctor's prescription for rest and calm during his chemotherapy. He drove himself back and forth from the clinic to the restaurant until the treatments were done. The intense pace kept his mind on food and off mortality.

Some regarded him as heroic, others as daft. I considered him lucky.

13

OLFACTORY ART

ONLY HUMANS DELIBERATELY create smells. People compose tantalizing perfumes and then package them in luminous glass bottles that are often etched to catch the light and create patterns as mesmerizing and as evocative as the scent within. Perfumers blend nature's essential oils together to enlarge the olfactory options, and then, in order to have a still larger selection to choose from at a much lower cost, develop artificial smells that can both ape the originals and multiply the possibilities of various aroma blends.

The hundred-and-seventy-billion-dollar fragrance industry employs some of the world's most perceptive noses. Those who aren't in the lab composing new scents are dreaming up ways to use them out in the world. Fragrances specifically designed for a particular ambiance and clientele are lavished on customers entering stores from Smith & Hawken to Victoria's Secret.

Cleaning products are a huge niche for artificial scents but also an endangered one. People don't clean the way they used to. A University of Michigan study found that from 1965 to 2005, time spent on household chores was slashed by 40 percent. To entice both men and women to focus more attention on the hunt for dust bunnies and the pursuit of the gleaming toilet bowl, fragrance specialists introduced new scents into products made by such companies as Procter & Gamble and Clorox. New scents are auditioned in testing rooms designed with a particular scent-savvy consumer in mind (people the industry call "scent seekers"). At Procter & Gamble, Anita's Kitchen belongs to a well-to-do homemaker, judging by the presence of Corian countertops and stainless steel appliances. A hidden camera records the reactions of test subjects recruited to experience the kitchen's smell. These reactions are deconstructed by scent specialists. A furrowed brow, said one, is "problematic." A twitching nose? Not good. That means the subject is asking, *What smells?*

The test subjects offer verbal opinions too, of course. More telling than what they say, though, is what they do — gestures, movement, facial expressions — in response to the scent, because the best scents toy with emotions without anyone's knowing it. If the test subject has to think about the odor, something's not right.

Even clean has its own smell. The trouble is, the smell of a clean space (devoid of odorizers, which clean freaks smell as a filth cover-up) isn't the same for everyone. Some say they want nothing but the smell of their own sweat to proclaim the housework done. They don't know their own noses. For these people, a faint whiff of bleach is the real closer. Clorox bleach comes in a variety of scents — it can smell like a spring meadow, or fresh linen, or lavender — but each of these is laced with a discernible whiff of ammonium chloride.

Some in the household-products industry are betting on ginger to carry the public into the next decade, just as the fragrance of fresh linen did the last. The growing brigade of anti-smell (and anti-housework) consumers may find the scent too edgy, though. Some specialists are putting their money on lavender vanilla, which is more soothing and thus better suited to a generation for whom housework may be redolent of unemployment.

What Febreze is to the hard-working middle class, Shalimar is to the ultra-rich. In terms of an olfactory art, perfume is even older than cuisine. In 2006, the *New York Times,* in a belated acknowledgment of that fact, hired its first-ever perfume critic. Better late than never. Perfume is part of everyday discourse in many European countries. Perfume journals are sold on newsstands in Paris, Rome, and Madrid. My friend Felip grew up near Barcelona, on the Mediterranean coast. A poet, he landed a university teaching job in St. Cloud, Minnesota. Only the name of the place is romantic; St. Cloud itself is as far removed from Barcelona as black is from white. It is cold and snowy seven months out of the year. An hour from the Twin Cities, it began as a farming outpost populated by practical Germans and is now infected by the worst kind of sprawl.

Picture a slight, openly gay man from southern Spain on this stark landscape. To remind himself of home he wears a bandanna around his neck that is doused in his favorite perfume, a strongly citrus scent that's light and fresh. He sprayed some on my wrist once. (This was just a few months before I lost my sense of smell.) The liquid ran down my arm and gave off a strong aroma of alcohol at first, out of which emerged smells that struck me as brilliant — as perfectly composed as a well-crafted musical phrase or a poem. I admired the beauty of the cut-glass bottle it came from, tall, a perfect oval shape, with a soft cushionlike dispenser that Felip took great delight in rubbing between his thumb and forefinger. He asked me if I knew how perfume reacted chemically with the skin. It had not occurred to me that it did. Does fragrance change with one's emotional state? Does it pick up the scent of nervousness that no doubt bubbles up through one's pores with perspiration? He was entranced with fragrance, Felip said, and more striking than his fascination was how unabashed he was about telling me so. Clearly such a statement would not be surprising in a place like Barcelona, not only because Barcelona is a cosmopolitan city but also because of its location, its temperate climate, and its age. The streets of Barcelona have been perfumed with strong man-made scents since ancient times. Food spices and sacred scents mingled in its open markets and in its perfumes. Over the centuries that scent acquired new smells from Italy, North Africa, Turkey, Persia, and the Far East, and Felip now equates it with home.

I told Felip that a family trip we took to Europe when I was a child had awakened me to smell. As I stood in the Old Quarter of Casablanca, our first stop, my nose was assaulted by smells of heavily spiced meats roasting on spits in the open-air market, live animals sweating and defecating in the street, incense and exotic perfumes, and fresh fruit rotting in the torpor of a Moroccan summer afternoon. The smells were set to music as strange and terrifying to me as the ubiquity of blinded eyes and toothless smiles, deformities, blemishes, and child beggars with bony arms and legs and stomachs swelled by malnutrition. As we moved north to Gibraltar, Messina, Naples, and Rome, I noticed that the still pervasive odors of poverty and heat were infused with delicious food smells. My nose was beginning to adjust. Taste helped it along. By the time we arrived in Venice, my fear of the unfamiliar had been vanquished and I found the mingled odors of urine and mildew clinging to the wet stones intoxicating. I thought Italian pasta was the best food on earth.

The olfactory shock that had had me in tears in the Casbah, begging to be put on a plane back to America, was gone. Europe had won over my impressionable young nose.

Up until the twentieth century, perfumes were made from essential oils distilled from plants; these organic molecules are called esters. In 1906 an English chemist named William Perkin received a knighthood for creating the world's first synthetic odor molecule: a test-tube-made copy of coumarin, which is a combination of plant extracts that has the pleasant fragrance of new-mown hay. After that, artificial scents took over from essential oils, and not just in perfumery. We have Perkin to thank for scented fabric softeners and deodorants as well as the new-car smell, without which taking home that shiny Honda just wouldn't be the same.

Perfumers still rely on petroleum products to extract fragrance from plants. Ko-Ichi Shiozawa, chief perfumer at Aveda (the company that made aromatherapy mainstream back in the eighties and nineties), is looking for alternative plant-based solvents. Shiozawa wants to wean the fragrance industry from its dependence on three thousand synthetic chemicals. His best effort so far is Yatra, an all-organic blend of Bulgarian lavender, soft Australian sandalwood (harvested by Aborigines), and South African geranium rose oils.

Another innovation in perfumery came in 2008. It didn't advance fragrance chemistry but it did take advantage of smell as a memory trigger. A New York clothing designer named Jessica Dunne smelled a market for scents that had a calming effect. Rather than relying on the usual relaxation herbs (lavender and so forth), Dunne decided to use old-fashioned flowery fragrances, the kind Dunne's grandmother Ellie had always worn. Just being around Ellie calmed Jessica down. She reasoned that lots of women her age (early thirties) had similar relationships with older women, or at least fantasies of days gone by that could be captured in the scents of an earlier generation. Why not bottle that?

Her perfume is a throwback to that simpler time, before the Me generation took over and designer fragrances made the wearer stand out in a crowd. An analogy to food might be apt: Dunne wanted to create a perfume that would be to Calvin Klein's Obsession what tuna casserole is to a spicy tuna roll. Ellie was a hit. "A brand is always a story well told" is how one perfume executive described the scent's appeal. Dunne was so encouraged by the sales of her quietly elegant (down to the bottle) product, priced at $180, she spun out a slightly more glamorous version of it, Ellie Nuit.

Dunne's success was a bright spot in an otherwise troubled time for the perfume industry. Fragrance is suffering from the combined effects of a poor economy and people's paranoia about artificial scents invading their personal spaces. With perfume banned along with tobacco smoke in many office buildings, people seem to be getting the erroneous message that fragrance is toxic. If the actual chemicals in a scent don't make you sick, the power of suggestion will.

Scientists expect that culture will condition people to have a universal squinched-up-nose response to the smell of burning leaves—a scent that not so long ago was a pleasant trip down an autumnal memory lane for many—in the same way that cigarette smoke can cause nausea in some individuals. Such a reaction can be psychogenic, brought on by research that shows that long-term exposure is harmful. The same conditioning persuades a person to enjoy the smell of body odor if it's his beloved's body odor and they're making love. This liking is turned off once he's no longer sexually aroused. How cultural conditioning and innate responses to inhaled chemicals collaborate to result in successful sexual partnerships is a complicated and mysterious process. Love and sex continue to mystify humans in spite of (because of?) our being obsessed with both. For whatever reasons, evolution has chosen to keep us in the dark. No one knows where the line is drawn between these two most basic instincts. Humans are eternally confounded by such questions as *Do I really love this person or is it just some mindless sexual attraction? Will the love* [or *the sex*] *last? How can I be trusted to choose the right mate?*

Poets have always known that the choice isn't really ours. Shakespeare never tired of this theme, and as a result we never tire of his plays. Composers too have long been inspired to concoct tales of love's labors lost (and found), none more bewitchingly than the nineteenth-century Italian Giuseppe Verdi.

14

A HISTORY OF THE
SENSUAL NOSE

O NE OF MY MANY resolutions after I became anosmic was to make better use of my ears. I'd been neglecting them. Scanning the Sunday *New York Times* for opera news in late March, I learned that Giuseppe Verdi's *Il Trovatore* was being considered for the Metropolitan Opera's 2009 season.

The most popular of Verdi's operas during his lifetime, *Il Trovatore* is seldom performed today. The trouble isn't the music but the ridiculous (to modern audiences) plot, which involves an olfactory misstep. The heroine, Leonora, is wooed in a dark garden by two singing brothers. One is her beloved Manrico and the other a notorious cad, the Count di Luna. "It's dark [and she's] working on smell, on the way her skin feels as the man gets close," artistic director David McVicar told the *Times* before opening night. "And here's [the singer playing Count di Luna] with his beautiful mane of silver hair," looking nothing like the singer playing Manrico. It should have been obvious to Leonora who was who. Yet she chooses the Count.

Verdi was born in 1813, more than a century before pheromones were discovered. To his contemporaries, it would have
been clear that Leonora had made a serious moral error by
trusting her nose. In the Age of Reason, olfaction was reviled
because of its connection to base animal instincts; the theory of
rationalism—"I think, therefore I am"—fit Catholic pedagogy
like a glove. By the nineteenth century, both concepts were being challenged by the Romantic movement, which celebrated
emotion and animal instinct. But their long partnership left
behind a Victorian culture deeply ambivalent and squeamish
about sex.

Olfaction hadn't always been held in such low regard.
Throughout much of human history, the primal sense was at
the top of the sense hierarchy precisely because it was anti-
intellectual and seemed to convey transcendent spiritual messages.

Old Testament writings often linked earthly love and spiritual passion through olfactory metaphors, and fragrant herbs
and oils remained central to the liturgy in early Christian
churches. In the ancient world, fragrance (and sex) was highly
esteemed. Cleopatra wouldn't have been seen, or rather smelled,
in public without first having her daily full-body, scented-oil
bath. Wealthy women wore hivelike appliances on their heads
designed to waft sexually alluring scents; the fragrances were
periodically replenished by servants. Aromatics were encased in
priceless receptacles and "served" during religious rites.

In the earliest religions, human foibles were not portrayed
in deeply negative ways. When people asked Greek and Ro-
man gods for favors, they usually wanted a good rain or some
golden elixir to fertilize the crops, not forgiveness for shameful
thoughts. Tribal peoples still use strong earthy smells in their

spiritual practices, which often involve drugs that induce smell-laden dreams they imbue with sacred meaning. Cultural anthropologist Constance Classen believes that dreams and odors have a lot in common: "Both are tangible and transitory. Both also can provide knowledge beyond that of the visible world, conveying essences hidden to the eye. In the modern West, odours are commonly thought to play very little, if any, role in dream-life. The Umeda, Ongee, Amahuaca, Desana and many other peoples know differently."

The message of Christianity changed as religion became an instrument of power. Everyone was a sinner, like it or not, and obedience to centralized papal authority offered the only hope of redemption. In the thirteenth century, mortification of the flesh promised rich rewards in the afterlife. As flesh was despised, so was smell. Christian ascetics practiced celibacy and slept on feces and mold to demonstrate their transcendence of the secular. The most extreme self-flagellators were canonized as saints. Poor hygiene and unsafe drinking water compounded smell's problems. Infectious diseases swept through the crowded cities of Europe. All these smells — from foul water to festering wounds — being both threatening and vile, became the scapegoats for illness caused by toxins as yet unknown. Convinced that the rancid smells themselves were the causes of illness, people took to wearing pungent scents to ward them off. Queen Elizabeth I is said to have worn a necklace of rotten apples, cinnamon, and cloves on the advice of the royal physicians.

The eighteenth-century naturalist Carolus Linnaeus, creator of the system for naming plants and animals by genus and species, failed in his attempt to create a workable taxonomy for odors, but he did guess correctly that sex was at the heart of every biological conundrum. When this put him at odds with the Church, he cloaked his suspicions in metaphor, implying that the flowers adorning the bridal bed were merely decorative, arranged by the "Creator" and "perfumed with so many scents, that the bridegroom and his bride might there celebrate their nuptials with so much the greater solemnity."

The fact that odors stubbornly resisted classification provided still more proof that smell was a bad actor—disorderly, anti-intellectual, and sexually promiscuous. Enlightenment philosophers quashed any hope that reason might apply a counterforce to religious teachings, at least regarding smell. I smell, therefore I am? Not likely in the eighteenth century. The opening passage in Patrick Süskind's *Perfume* describes eighteenth-century France this way:

> In the period of which we speak, there reigned in the cities a stench barely conceivable to us modern men and women. The streets stank of manure, the courtyards of urine, the stairwells stank of moldering wood and rat droppings, the kitchens of spoiled cabbage and mutton fat; the unaired parlors stank of stale dust, the bedrooms of greasy sheets, damp featherbeds, and the pungently sweet aroma of chamber pots.

The odors conspired in the public imagination to create a reputation for smell that was

> straight out of the darkest days of paganism, when people still lived like beasts, possessing no keenness of the eye, incapable of distinguishing colors, but presuming to be able to smell blood, to scent the difference between friend and foe, to be smelled out by cannibal giants and werewolves and the Furies, all the while offering their ghastly gods stinking, smoking burnt sacrifices. How repulsive!

Educated elites responded as they always did in these wars: they said one thing and did another. Few of the pampered ladies of Versailles fretted over the sinfulness of their ways; what woman had time to think about that when she had a three-foot-high hairdo to keep from toppling into her glazed strawberries and a décolletage that needed constant freshening with sultry scents? In the eighteenth century, fashionable women anointed the cleavage with elaborate concoctions of their own devising. (One, called angel water, combined orange flower water, rose water, myrtle water, musk, and ambergris.) The breasts were elevated until only the nipples were concealed, and between the "hillocks," as the poet and writer Diane Ackerman termed this vision of heaving pulchritude in *A Natural History of Love,* were placed cut gemstones, just in case the scents and the push-up bra weren't enough to "waylay the eye" or "lure a wayward nose."

Europe eventually cleaned up its act. Odor was banished. By the Victorian age, legs were called "limbs," and neither smell nor sex was discussed openly. Women were tightly supervised and fully covered, their tiny waists cinched in with death-defying corsets that sent a mixed message. Smell was in the air but not in polite conversation. A researcher who did an informal literature search on the role of smell in fiction discovered (not to her surprise) what she regarded as a regrettable decline in olfactory references that began in the nineteenth century and persists today. She links this "deodorizing" to our modern obsession with tidiness. A clean house (and appearance) reflects a clean, chaste, and virtuous mind.

One who benefited from his own intuitive grasp of the downside of dualistic thinking was the neurologist-turned-analyst Sigmund Freud. Based on his work with female patients, Freud concluded that a fear of sex (manifested by repressed sexual desires and experiences) caused mental illness. Both sexes had the fear, he believed, but women were especially vulnerable. When Emma Eckstein, who happened to be extremely beautiful (and blessed with a prominent nose), sought his counsel, Freud saw an opportunity to test a radical notion involving mind-body connections that a close friend and colleague had developed.

The colleague, Wilhelm Fliess, was a medical doctor specializing in maladies of the ear, nose, and throat. Fliess had come up with a theory that the female nose controlled the sexual response and was linked to the genitals by some unknown pathway.

Eckstein had sought Freud's help for what a modern gynecologist would treat as a routine menstrual problem. Freud's suspicion of an underlying psychic disturbance was fueled by her revelation to him that she'd been sexually abused as a child. He made note to Fliess of her nose. Its size and shape suggested that she might be suffering from what Fliess called "nasal reflex neurosis." Fliess diagnosed the patient with the disorder; a misshapen turbinate bone in the nose was, he claimed, messing with the messages sent from the young woman's nose to her vagina. He removed the bone.

Did corrective surgery on her nose cure her? Alas, it did not. Not only did she fail to improve, but several weeks later she nearly bled to death due to Fliess's postoperative negligence (he had forgotten to remove nearly half a meter of surgical gauze). Freud was faced with a dilemma that some believe determined the future course of his career. The dilemma was this: he could disavow not only his friend Fliess but also his own budding theories about sexual repression — or he could blame the patient. Freud chose the latter option, accusing the young woman of inventing the childhood-abuse story and therefore causing the unfortunate results of the surgery. "Bleeding hysteria" was how Fliess and Freud later referred to the "condition" that almost killed Emma Eckstein. Given the ignorance about sexual function and the powerlessness of women at the time, it is hardly surprising that Freud's subterfuge worked.

To me sex has always seemed like a disconnect of a different sort: male versus female sexual appetites. I believe this discrepancy, while not universal, has to do with the two genders' different roles in species survival. Freud would have been bored silly by the simplicity of my argument that the male is a biological hard-ass ever looking outward to protect the family and find potential mates and that the female is hard-wired to keep her babies alive through behaviors now lumped together in the word *nurturing*.

Gloria Steinem might also object. But hear me out, fellow feminists. Women typically have superior social skills (higher emotional intelligence) when compared to men, and men are more likely to follow logical reasoning to idiotic extremes. Freud was a supremely logical fellow, after all. He possessed the classic male blend of logic worship, derring-do, and will to power. In his zeal to probe new scientific territory, he almost killed one of his patients. To save his reputation, he covered it up.

I forgive him, even though his ideas would strike the average person today as far more hysterical than Emma Eckstein's menstrual complaints or her claims of child abuse; in fact, they're borderline evil. But Freud was apparently just as skilled at repressing inconvenient truths as he believed many of his patients were. Someone has to go out on a limb to give us the benefit of hindsight. Science is all about risk-taking and testing hypotheses. I'm frankly tickled pink that more and more women are behaving like men. By that I mean, taking intellectual risks and betting the henhouse on a hunch.

I'll discuss in more detail how smell helped form the biological underpinnings of gender behavior and sexual attraction in the next chapter. This one ends with a caveat.

I confess that I grew up Catholic. Confessing is practically hard-wired in people like me. Sex is not. I was an *extreme* Catholic. I attended mass daily, prayed myself to sleep at night, kept a candlelit shrine to Saint Dominic Savio in my closet, knew the catechism by heart, and took quite literally those subliminal warnings (conveyed through the use of pictures showing souls in various shades of black) that the soul actually *did* darken (and Jesus wept) in response to my moral depravity. Never mind the physical act of sex. It was the thinking about it that was so pernicious — and of course unavoidable. I was doomed.

Enjoyment of life's sensual pleasures still doesn't come easily to me. True, a garden is the ultimate self-indulgence, rich in sensory delights. But my addiction is redeemed (in my mind) by muscle tears, thorn pricks, skin lacerations, and sweat. As to sex, I'd rather talk about it, parse it, analyze it, and deconstruct it than do it. For me, one of smell's best features had been that it caught me unaware, putting me in a romantic (i.e., receptive) mood in the bedroom. Cam's after-shave, a product called Eau Sauvage that his French brother-in-law had turned him on to decades ago, always used to have that effect on me. And what better than smell to subtly alter whatever mood my reading had put me in and send me unthinkingly over to Cam's side of the bed? The scent of his skin and hair — rosemary and Dial soap, the manlier fragrance of Barbasol Beard Buster shaving cream, and, later, the new after-shave he switched to after Eau Sauvage was discontinued in the States—used to beckon me there without my even knowing it. Simply put, body odor and musky scents signal sex, while flowers and perfumes and the smell of a person's sweater spell romance.

When I lost my sense of smell, all those sensory cues vanished. Deprived of my husband's familiar (and to me extremely attractive) scent, I sometimes forgot he was in bed beside me.

15

THE EROTIC NOSE

How could Verdi's tragic heroine have been fooled by a fragrance? Quite easily, in fact. Many biologists who study the brain are becoming convinced that pheromones are processed in the human olfactory system even though they don't necessarily have smells. Classified as ecto-hormones (*ecto* means "external" in Latin), pheromones are secreted and convey biological messages between individuals of the same species. These signals can affect both hormone levels and behavior. Just as a hawk can't interpret the unflattering things being said about it by the squirrel cowering in its shadow, the pheromone signals that lure a male pig to a fertile female pig don't have any effect on, say, monkeys. That's a good thing for both species, biologically speaking.

The term blends the Greek words *pherein* (to transport) and *hormon* (to stimulate). It was dreamed up in 1959 to denote the chemical released by female silkworms to attract males. Discovery soon led to invention: bombykol, a commercially produced version of the silkworm pheromone, works the way all pheromones do, and farmers used it to lure problem insects to deadly traps. Tiny amounts did the trick. Bombykol was safer and cheaper than spraying with the standard repertoire of chemical pesticides.

We know that insects respond to pheromones; people are another matter. But if human pheromones are delivered as smells to the olfactory cortex, which some scientists are convinced they are, fascinating possibilities open up. A woman is more likely than a man to take note of a passing scent; she engages her thinking brain in identifying and labeling it, and it's stored in her limbic system for future reference. (Men, meanwhile, are storing football scores.) Like female silkworms, female humans are the aggressors in sex, in that they wield more influence in deciding who will mate with whom. Males are less discriminating for good reason: they are less involved in the outcome of sex, children.

Girls pick up language earlier than boys do, and most women can outtalk their boyfriends or husbands handily if the measure of success is words spoken per minute. That women tend to be the talkative sex seems at first glance illogical, since language has always been indifferent, even oblivious, to smell. With females' relatively big vocabularies, shouldn't the gender at least have come up with a word for the smell of an old sock? Some scientists believe that women are more verbal than men because the part of the female brain that processes language is closer to the amygdala than the corresponding area is in the male brain. Women are also more likely to suffer from depression than men. (However, homosexual males differ from heterosexual males; their brains respond as women's do to emotional and olfactory stimuli.) Women are better at telling smells apart and labeling them. Intuition suggests that this is because of the female's primary caretaking role: Dad goes hunting while Mom stays home with the kids. Women were the earliest cooks and gardeners. Evolutionary biologists are convinced that the males of almost all species are the more flamboyantly adorned because women are the aggressors in sex, and along with the males' fancy costumes are pungent odors, though whether pheromones (if they exist) smell is unknown.

The male-female discrepancy in language, empathy, and smell IQ becomes most noticeable during puberty and keeps expanding after that. Could those adolescent hormones possibly be pheromones?

Sometimes described as compounds, sometimes as steroids, sometimes as hormones, pheromones are that mysterious something in the urine and feces that dogs use to mark their territory. They are what pig farmers use to ignite sexual passion in female hogs. In people, they are distinguished from sex hormones, such as testosterone, which turns sexual arousal on and off by its unique and extremely subtle communication skills. However, testosterone does not direct a man to a woman whose genetic makeup is sufficiently unlike his own to favor healthy offspring. Pheromones do. That is, if they actually exist.

A famous study found that a wife can almost always pick her husband's T-shirt out of a pile of men's T-shirts with her eyes closed. If a man is "the one," even the smell of motor oil on his hands can arouse romantic feelings; the same mechanism explains why the smell of methyl mercaptan, the substance added to scentless natural gas, makes people anxious. Smell skillfully weaves together the sweet fragrances of romance and the more assertive odors of the Darwinian imperative (a.k.a. lust) to create the complex tapestry of impulse and emotion called love.

This is where the discussion bogs down. No one is speculating that methyl mercaptan is a pheromone. How does one parse pheromonal cause and effect when no one is even sure that human pheromones exist or, if they do, whether smell has anything to do with their delivery? All researchers definitely know is that these chemicals control the sexual behavior of silkworm moths, whose larvae are the silkworms that turn out silk. Smell has a far subtler role in humans' sexual activity than in bugs'.

Yet even the Victorians whispered of certain olfactory aphrodisiacs, and these days pheromones are the talk of the Internet. One website hawking a product seductively named Pherlure cologne promised that armed with the cologne, the human male could, just like pheromone-producing members of the animal kingdom, "arouse the female sex glands, heighten sensual responses and awaken her appetite for sex." That products such as this may not actually contain pheromones doesn't mean they're entirely worthless. Sexy smells can be borderline repulsive, and the smell of urine, often detected in body odor, is also in musk. One man's musk is another man's urine. Literally.

Which is probably why my husband did not find me irresistible when I picked him up at the airport after I'd slept in the same clothes for three days and not bathed. Cam stuffed his bags in the trunk, slid into the passenger seat, said hi, mentioned that he was fighting off an earache, and kissed me on the cheek, explaining that whatever he had might be contagious. "I wouldn't want you to wake up one morning stone-deaf." Then he stopped in midembrace and, lifting his hand to cover his nose, said, "Man, we gotta talk. Have you showered lately?"

My immediate response was the predictable spike in blood pressure — mortification at full throttle. It was the smoking-saucepan incident all over again. Then it dawned on me that Cam had probably been dying to discuss my body odor for weeks but had held his tongue.

So why is science interested in pheromones? Legitimate fragrance manufacturers don't have squadrons of chemists hunting down the next-generation Pherlure perfume. Even if such chemicals are at work in human reproduction, odds are they operate far under the radar. Which is why biologists are interested in what they do. If pheromones do in fact arrange the best biological marriages, they could answer puzzling questions about evolution. We know that Mother Nature seems to want to pair off the fittest individuals, sometimes only for the split second it takes for one to impregnate the other, sometimes for life. The duration and quality of the male-female hookup depends of course on what's best for the next generation as a whole, not for the couple, as Leonora's plight certainly attests.

Nonetheless, there is evidence in the animal world that in individuals who mate for life, happy marriages beget healthy kids. One researcher has traced the monogamous instinct of a particular type of prairie vole to a pair of brain hormones. These hormones (or pheromones — the words are sometimes used interchangeably) show up not only in circuits of nest-building mammals but also in those of humans. They may be involved in love itself. In 2008 scientists tested the theory of this so-called monogamy gene on humans. While the results were inconclusive, a significant correlation did emerge between responses to certain pheromones and the males' marital track records.

The Wisconsin National Primate Research Center at the University of Wisconsin–Madison announced in the fall of 2008 a striking discovery involving marmosets. These small New World monkeys live in tight-knit nuclear families, a model exceedingly rare in nature. The dads actually stick around *and* pitch in with the housework. Those dads, it turns out, are given a sort of biological anti-Viagra when newborns come. This is in the form of a sniff of a chemical — call it a pheromone — the infant marmoset secretes only at this vulnerable time, when Mom could really use some help. The chemical causes the male's testosterone levels to drop off dramatically, resulting in a distinct personality change. He behaves less like Marlon Brando in *A Streetcar Named Desire* and more like Dagwood Bumstead in the *Blondie* cartoon strip.

Why do marmoset dads get this unusual postnatal shock treatment? Scientists think it's because marmoset babies are a handful. Like human babies, they are born big, weighing in at 10 percent of their adult body weight. Plus, they come in pairs. Of particular interest to the researchers is that the marmoset males studied in their natural habitat were as aggressive as usual in protecting their turf and newborns. Apparently, in emergency situations, the male receives a jolt of testosterone. Afterward, just one sniff of the newborn is all it takes to bring the level back down.

Catherine Dulac, a Harvard molecular biologist who specializes in the genetics of smell, conducted an ingenious experiment on female mice who'd just had litters of young pups. These weren't ordinary moms. Dulac had tampered with certain ion channels, blocking the pathway of a pheromone that represses male behaviors and encourages female behaviors. This pheromone kicks in at specific times, such as when a mouse mother is supposed to be paying attention to her newborns. The result of blocking the pheromone's influence? Gender-confused creatures who brazenly mounted their mates (instead of passively waiting to be mounted) and who ignored their young. The mice were physiologically unchanged — all of their sexual equipment remained female. Yet they stopped *acting* female, and even failed to deliver the licks and cuddling believed to trigger responses essential to baby-mouse development. The babies of the altered female mice failed to thrive and eventually died.

Dulac called the results "flabbergasting" and proof positive that female mice have the same neural circuitry as males. The only thing separating the women from the men in the mouse world, she concluded, was a pheromone—sent from a female's children or her mate or both—that repressed male behavior and activated female behavior. When the researchers altered a gene and blocked the effects of that pheromone, the female mouse lost her ability to be a good mom.

Apparently, human babies send pheromonal messages too. It's a well-known fact that a blindfolded mom can tell her own newborn from other babies in the nursery just by the infant's scent. Researchers have identified a patch of skin on top of a baby's head that has cells that emit this powerfully bonding one-of-a-kind odor. Babies also seem to have an uncanny ability to pick their own mothers out of a crowd.

Next questions: How is the pheromone signal processed? How does it enter the brain and what gene decodes it and where does it go? An organ that Richard Axel calls "the erotic nose" —this is the vomeronasal organ (VNO)—is widely believed to be in charge of processing pheromones in most animals. The VNO is not the same thing as the olfactory organ, though it is located in the nasal cavity and operates, Catherine Dulac suspects, in concert with the smell system. The VNO is a chemosensory organ, meaning that it detects and responds to certain chemical stimuli, such as pheromones. In Dulac's model, the vomeronasal pathway "serves as a switch that represses male behavior while promoting female behavior. While male and female bodies are strikingly different physiologically, it appears the same cannot be said for the brain."

Dulac and her colleagues have shown that the mouse olfactory epithelium and the VNO together play a critical role in signaling for sex. Their job isn't to tell male from female but to trigger sexual behaviors. Dulac's hunch is that humans are no different from rodents in this respect; the only variation is that in humans, the olfactory system alone processes such signals and does so without any connection to consciousness. Others disagree. Some scientists argue that if the VNO is the organ that handles pheromone signals, it's unlikely that humans have pheromones. Why? A human doesn't have much of a VNO. It atrophies before birth. While recent studies indicate the VNO may be operational even in its truncated form, Dulac thinks the whole issue is a red herring. So what if we don't process pheromones the way other animals do? Couldn't they be processed along the same pathway as smell?

This brings us back to Leonora. According to Dulac, a woman doesn't *know* that a man's scent is attracting her because there is no direct neuronal link between the olfactory cells assigned to the task and the higher brain. As Dulac explains, "Our data contradict the established notion that VNO activity is required for the initiation of male-female mating behavior." This, she says, is the key difference between pheromones and the odors commonly mistaken for them, and it's why she doesn't think pheromone responses are the result of training. She argues that while odor signals are distributed throughout the brain, even up to the cortex, pheromones take the express route to organs in the limbic system that prepare the body for sex through changes in the sweat glands, breathing, heart rate, and so on.

In 2006 Linda Buck wrote that VNO receptors were functionally distinct from odorant receptors and that they appeared "to be associated with the detection of social cues." She reported finding a second class of pheromone-linked receptors in the olfactory epithelium that could be central to the reproductive process in mice. This new class of chemosensory receptors suggests that humans too may respond to certain volatile compounds outside of the standard repertoire.

In a joint study published in *Nature* in 2007, Rockefeller University and Duke University researchers announced they'd found genetic variations in a gene encoding for an odor receptor that detected androstenone and androstadienone (present in human males' sweat and urine). The different genetic variations in the receptors resulted in different perceptions of the hormones' smells. This was a first. Never before had biologists made a direct link between genetic variability in olfactory receptor genes and differences in the perception of a smell. Hanyi Zhuang, a graduate student who worked on the project, thinks the scent of androstenone helps a female evaluate potential mates. How a woman responds to a man is determined by whether the chemical jump-starts the physiological changes in mood that constitute what we call, curiously enough, chemistry.

Writing in the September 2007 issue of *Scientific American,* in the article "The Scent of a Man," Nikhil Swaminathan explained that the presence of a certain odor receptor in a woman determines if she'll detect the scent, and that genetic variations in that receptor determine how the scent is interpreted.

The fact that the human VNO atrophies in utero may explain why humans are relatively poor sniffers when compared to other animals but are good at seeing. (In fact, we are excellent at smell discrimination; a human can detect nuances that escape even a dog's nose. That is, if the human can detect the smell in the first place, which is where our species falls hopelessly short.) Either we needed pheromones incorporated into the system with some cognitive input, or we didn't need them at all. Pheromones might enter the human brain through the nose, or possibly the skin. If they don't enter at all, this may be because the genes for sex have been altered in humans to better accommodate monogamous lifestyles. A person can't afford to have sex on impulse, as is inevitable when sexual response is controlled by chemicals that don't have a thought process attached. Some other animals don't *have* a thought process; they can have sex anytime with anyone because their children are born relatively self-sufficient. However, in many species, pheromones emit signals that tell prospective male suitors who come calling on a female with newborns: *Now is not the time. This one's got her hands full. Take a rain check.*

Richard Dawkins suggests that evolution may have a hand in the phenomenon scientists call the Bruce effect. Bruce is the guy (male mouse, actually) who attempts to impregnate a female who already has someone else's babies on the way. Rather than give birth to babies belonging to a different suitor, she aborts. The theory is that the new male mouse — Bruce — secretes a chemical that causes the pregnancy disruption, which makes the female available for Bruce to impregnate.

While marmosets definitely have a working VNO, Old World primates do not. The University of Chicago's Yoav Gilad and others speculate that when evolution decided to abbreviate the life span of the human VNO (back when we were primates), it simply rerouted pheromone pathways through the nose, jiggering the genes and receptors there so they'd recognize the pheromones. Some researchers think that although the fetus's vomeronasal nerve is gone by the second trimester of its development, the nerve function is still active . . . somewhere. Either way, a consensus seems to be forming that humans do process pheromones somehow; if not in the usual animal-world place (the VNO), then through the nose. And the pheromones' under-the-radar influence may actually strengthen their power over human behavior: it's hard to resist what you're not aware of.

University of Pennsylvania psychologist and smell expert Richard Doty doesn't believe that pheromones exist in either people or animals. In bugs, yes. Like humans, insects lack a VNO. Pheromones are processed through the olfactory system—the antennae—and are all about sex. (For example, for years trees in Beijing had been ravaged and defoliated by moths. To make the trees leaf out for the 2008 Olympics, the Chinese hired a Finnish entomologist and pheromone expert; in the lab, the scientist used pheromones to accelerate mating in leaf-chomping moths, then he infected the moths' larvae with lethal parasites. The infected larvae made cocoons, which were placed in trees all over Beijing. When those cocoons hatched, parasites emerged and swarmed through the trees, killing all the leaf-eating moths. The trees leafed out just in time to ensure a green Olympics.)

In *The Great Pheromone Myth*, Doty compares pheromones to Snarks, an analogy inspired by the Lewis Carroll epic poem *The Hunting of the Snark*, in which the poet makes fun of how far humans will go to explain the unexplainable. In the poem, the Snarks' entirely theoretical existence became as real to the theorizer as a certain birthday suit was to a certain emperor. To debunk pheromone theory, Doty pointed to the perfectly normal sex lives of congenital anosmics. He believes behaviors currently being attributed to the invisible hand of pheromones are just as likely to come from conditioning. He notes that cultural anthropologists consider kissing more olfactory than pheromonal, and that kissing is sexual only in certain cultures where that response has been conditioned by society.

Modern society could just as easily consider touching noses sexual, as the Eskimos do. Apes and chimps kiss as a form of peacemaking; sex has nothing to do with it. Mongolian fathers don't kiss their sons; they smell their heads. Charles Darwin described a tribe called the Malay in which women greeted their male kin by squatting on their haunches and turning their noses skyward; the men responded by vigorously rubbing the women's noses. Darwin regarded the ritual as a sort of hand-shake. Even that rite of greeting has an olfactory component; it allows a transfer of scent when the hands touch the nose. This intimacy is also seen in hand kissing, an outgrowth of hand sniffing, and is related to dogs' habit of licking their masters on the mouth to savor their identity.

Doty doubts that pheromones are the secret weapon Mother Nature deploys to ensure species survival; he doesn't believe that pheromones can direct a woman to a man whose immune system offsets deficiencies in her own; or that they can lure a man to a woman whose high hip-to-waist ratio tells him she's able to carry a baby to term; or that they can help a female nose out a guy whose rugged good looks (signifying a high testoster-one level) predict rugged children.

Doty argues that training (the subconscious kind), not pher-omones, can easily explain certain biological phenomena, such as why marriage partners in a religious sect called the Hutter-ites, an insular religious community with a limited gene pool, tend to have dissimilar human leukocyte antigens, a diver-sity that's crucial in fighting pathogens. Are pheromones the matchmaker, the mysterious X factor that protects the Hutter-ite gene pool from the ill effects of inbreeding? Scientists are only beginning to understand the relationship between genet-ics and the immune system—for example, why some who pos-sess the gene for a form of Parkinson's disease develop the disorder while others who have the gene do not (could exposure to an inhaled toxin act as a gene trigger?).

Experiments continue to show that something in the human chemical makeup, whether or not it's pheromones, is at work to keep our species alive and mating well. Swiss researchers found that the female subjects were consistently drawn to T-shirts belonging to men whose leukocyte antigens differed from their own. At UC–Berkeley, female undergraduates showed measurable signs of arousal — physiological and psychological changes —when they sniffed androstadienone, a steroid found in male sweat. Psychology fellow Claire Wyart conducted the Berkeley study. She admits that the strongest argument against human pheromones is that members of our species do not exhibit stereotypical behavior when a particular substance is detected.

Using a brain-scanning technique called positron emission tomography, Ivanka Savic, a neuroscientist at the Stockholm Brain Institute, exposed subjects to androstadienone and then tracked their brains' electrical activity. In the brains of straight women and gay men, the electrical activity increased, indicating sexual arousal; in the brains of lesbian and straight-male subjects, no heightened activity was noted; to them, the steroid was just another common odor. However, when that group was exposed to a female steroid, they responded in the identical way the straight females and gay males had to androstadienone. Again, though, we don't know whether these responses are hard-wired or learned.

If androstadienone is the male hormone that drives females wild, what (if anything) lures the men? Possibly the steroid estratetraenol. Claire Wyart and her colleagues are quick to fend off another wave of online "pheromune" ads by noting that there is no evidence that these compounds have the same "subliminal or instinctual" influence on mating in humans as they do in insects. That's because pheromones don't follow the same rules (i.e., pathways) that smells do.

In a groundbreaking 1971 study that launched this line of research, Martha McClintock showed that undergraduate women living together almost always developed synchronized menstrual cycles. She believed a female pheromone triggered the effect. But why? Some think "primitive" man had multiple sex partners, and so the synchronous cycles improved the odds that he'd mate with a fertile woman. Or maybe the synchronized menses are connected to the monogamy gene. In close-knit communities, adulterous affairs could put children at risk of neglect. That women's cycles synchronize only if they ovulate reinforces this notion. Birth control pills eliminate ovulation (and perhaps the influence of pheromones).

Could these so-called pheromonal responses actually be learned behaviors? In the case of androstadienone, perhaps the women were just responding to the musky smell of the androstadienone, not to the compound itself. (The researcher knew as well as Doty that if you put good sex together with the smell of, say, cat pee, eventually even cat pee can be an aphrodisiac.) In a 2002 study, Martha McClintock investigated this theory. She tested androstadienone and two other substances — androstenol, a musky-smelling steroid compound, and muscone, a musky-smelling *non*steroidal compound. She used the strong smell of clove oil to mask the muskiness of all three compounds. The women in her study showed elevated moods in response to the androstadienone — and *only* the androstadienone. More proof that pheromones exist, but hardly the smoking gun. As McClintock herself wrote, it "is yet to be determined whether humans exude concentrations into the air adequate for social communication or process this chemical information within natural social contexts."

Richard Doty's central question remains unanswered, but the evidence seems to be mounting: something in our genes responds to something in the air, and we as individuals have no control over it.

16

THE LANGUAGE OF SMELL

M Y LOVE LIFE was the least of my worries. Cam had put up with me this long and he seemed unlikely to bolt now. But was anosmia eroding my language skills? My husband suspected that this worry was just my latest excuse for turning down paying work. He thought my language skills were fine. I was talking more than ever, and about some pretty complicated ideas. My vocabulary seemed to be growing, not shrinking. I actually knew what a hippocampus was. Maybe *he* was the one with the brain problem; he couldn't remember what it was from one science lecture to the next.

"The hippocampus stores memories," I reminded him for the fifteenth time. "It's located in the limbic system, right behind the eyes, and is shaped like a sea horse."

"Frankly," he said, "I wish you'd take your vocabulary up to the attic and write so I can watch some basketball."

I was beginning to picture my hippocampus as a sea horse riddled with holes. Those tip-of-the-tongue moments when a word wouldn't come were getting more frequent, or so it seemed. That the brain is not wired to identify odors out of context is another result of this sense's great age; it explains why even a person with a normal nose typically guesses right just one in ten times when asked to name an unusual smell. (Common odors are identified correctly about half the time.) But unlike tip-of-the-tongue phenomena, a tip-of-the-nose experience allows no verbal information. Smell is really a recognition system, not a naming system. Trygg Engen explains in his book *Odor Sensation and Memory,* "Sight and hearing are more cognitive. Smell is emotional and does not have to be rational."

One morning I came downstairs after Cam had left for work and found an article he'd torn from the *Wall Street Journal* waiting for me on the kitchen table. He'd scribbled *interesting* across the top. Knowing my husband, *amusing* is what he really meant. He was trying to stop my morbid ruminating on the crippling consequences of anosmia. He was daring me to laugh. The reporter had been traveling in Scotland and had interviewed some people in the whisky business. He discovered that professional Scotch whisky tasters have an unusual talent for dreaming up quirky descriptions of the way whiskies smell. Though *taster* is the official name, and sniffing is what they actually do, they truly shine in the word department. One of them described an especially fine Scotch as "smoky and peppery with a hint of mechanics' rags." The reporter concluded that the whisky seemed to "elicit responses more revealing than any ink blot ever did."

And as long as the description doesn't get put on a billboard, if a whisky tastes like Bazooka bubblegum, why not say so? The nose—that is, the aroma—of one whisky was described as reminiscent of cotton candy and lipstick. Another had a "smoky, oily quality," and its smell was compared to both bacon fat cooking in a tractor shed and heather in a closet. "Smoldering slag heaps with brown sauce" from a chip shop, anyone? Or how about a combination of carnations, roses, and "the rubberiness of school erasers"?

Smells are like love: Irrational. Elusive. Just the thing to get the similes flowing, to spark the imagination. In 2000, a team of French researchers found that "the world of smells is difficult to pin down in words" for what they felt was one of two reasons: either the ancient olfactory system is encumbered by its weak neural connections to the "youngest language brain structures," or verbal descriptions of smells simply aren't necessary. Maybe because smelling predates self-awareness and speech, smell doesn't need words. And, I hoped, words don't need smell.

I also hoped that the poet and essayist Andrei Codrescu was being facetious when he remarked on National Public Radio that the human "smell-descriptive vocabulary" became "more impoverished the more our language developed." Codrescu says the first time he took his dog to New Orleans, the "layers of funk made her so excited she didn't sleep the whole time. Later she told me it was an encyclopedia of smells you'd need a million years of brilliant poets to find a language for. She told me this in dog, with her nostrils trembling. She chooses not to speak human, and from an olfactory point of view, I can see why."

Purported to be a purebred bichon frisé and sold for eighty bucks in a Burger King parking lot, our dog Mel was a last-minute Valentine's gift from a boyfriend to my daughter Alex. The dog turned out to be a rat terrier with a soupçon of bichon frisé evident in his affectionate nature and his upward-curling tail. He's hard-wired with a single idée fixe (that would be the bichon part talking): *I don't want to go back to that puppy farm.*

Mel's circus act is one of the more impressive results of

Alex's six-figure college education. Over the span of two years spent in a tiny campus apartment crammed with homesick, dog-worshiping females, Mel developed a routine that grew to include not just the obligatory *shake* and *sit* but also *roll over* (three times), *stand* (for half a minute, paws flailing), and *take a bow* (head on front paws, rump high in the air, tail wagging). The first time Alex put Mel through his paces for me, I was astounded by what he would do for a Milk-Bone.

When Mel and I talk to each other, I speak English, so it's a fairly one-sided conversation. But when we smell each other, I'm the idiot and he comes away enlightened. Mel can't tell steak from hamburger, but he knows if I've been gardening, driving, or hanging out with Cam. He knows if I'm nervous or just need a bath. He knows if I'm angry, hungry, happy. What do I know back? My nose can tell me if he's been rolling in deer scat.

Mel can smell me coming a block away and runs around to the gate to greet me with his high-pitched whine-bark. He licks my face to absorb and revel in its scent, then licks himself afterward to transfer the scent of safety and comfort so that the smell will be there when I'm not. He loves me, but my personality—my sense of humor—is lost on Mel, and he has to put up with my insensitivity regarding many of his emotional needs. That's all due, at least in part, to the most glaring difference between us: the language barrier.

Dogs talk to each other all the time with no trouble at all. They don't even need to be in the same vicinity. Better if they're not, in fact. Their habit of marking territory is as close to language as they get, and it is a communication system derived from smell. Scientists are just as inventive in developing theories about the method behind what looks like madness as those whisky tasters are in coming up with word pictures that capture the scent of a Scotch.

Mel becomes quite the social butterfly when we're out walking, darting from tree to telephone pole, picking up messages and leaving his own, livelier in his "conversations" than a teenager on Facebook. Some friends he's never laid eyes on. Makes no difference. His nose tells him everything he needs to know. Male or female? Sweet-tempered or a possible threat? Longtime resident or new in the neighborhood? On some level of consciousness accessible only to canines, he knows. If I forget to take him out, he lifts the latch on the gate with his nose and sets off on his own. He never goes very far or stays away long. Maybe we have his early-childhood trauma to thank for that.

Marking is the canine equivalent of talking. But dogs get right to the point. They look out for those with whom they live, regardless of whether home is a cardboard box or a cozy bungalow with a fireplace and central air. The following dumbed-down distillation of several academic treatises on marking may still sound like jargon, but it goes beyond the image of a dog lifting its leg to convey the sophistication of canine communication: A nonresident dog adds up the number of identical marks (in piles of poop or on pee-streaked fire hydrants and so on) to find out who's in charge in this neck of the woods. It's that guy with the weak bladder. I have noticed that Mel lifts his leg more often the closer he gets to home. Dogs also become much more aggressive when they're on their own turf. Uh-oh, dogfight. (I've noticed that too.) Dogs have pheromone-producing glands around the anus. Sniffing another dog's behind tells the intruder that this is indeed the resident—the dog whose presence was heralded by the smell fingerprint left on its feces. It also works the other way around. The remembered odor of that pile of poop lets intruders know if they're trespassing. All they have to do when another dog nears is get a good whiff of its rear end.

Zoologists describe marking as a way to give intruders a heads-up should they happen to run into the resident canine (the dog who lives there). Marking says, *Back off.* Miraculously, the intruder obeys. By keeping "agonistic encounters" to a minimum, marking protects the entire species from extinction.

Lest we get all teary-eyed over what lengths our pets are willing to go to protect us, consider what is going through their minds as they sniff a bush damp with pee. They're not thinking about us at all but responding to a complicated message that has been condensed into a split-second call to action. Scent-gland secretions, feces, and saliva impart vital clues as to the whereabouts, age, gender, social status, and emotional state of the dog involved. Intruders compare new scents with remembered images of a previously sniffed one. Only the most recent scent image left by a dog is meaningful to the intended receiver. A dog that can defend its turf long enough to mark it comprehensively is sending a strong message, namely, that it's quite the athletic brute. The intruder knows that of the two of them, the resident is more motivated to fight. This is his home. He knows it cold. He has more to lose than some unwary passerby, which is why intruders almost always withdraw. Sorry, must run.

Why do dogs roll in their own feces? This isn't to gross out human companions but to let intruders know who's been marking this territory — the resident. Likewise, a dog that bolts out the door and barks at passing dogs is making his scent available for matching. *Yeah, I'm* that *dog,* he's saying. And when you're out walking your dog and he sniffs another dog's poop, it's not because you forgot to feed him. He needs to know for sure that the barking dog he's just passed lives here and means business.

The ancient Komodo dragon doesn't behave all that differently from Mel when it comes to marking. It deposits fecal pellets to mark its territory; a young, intruding Komodo will sample one (by nose and by mouth), and if it determines that the owner is older, meaner, manlier, and more invested in the turf than the younger one is, the new Komodo will make an appeasement display.

Intrigued by these creatures, I went hunting online for footage showing a live Komodo in action. Forget Mel's humanlike traits — the giant lizard's similarities to humans are even more striking, though in an extremely macabre way, I admit. A short video showed a Komodo on the hunt: the slow, patient pursuit of prey first detected by the forked tongue stabbing the ground for scent; the gleam in the eye when a fat water buffalo sensed that it was in danger; the saliva dripping like a bead of wax from the Komodo's immense jaw as the lizard anticipated the coming feast; the kill itself — a single chomp delivered to a rear leg. Then, creepiest of all, the Komodo's two-day death watch as the buffalo weakened and dropped to its knees, then to its belly, and finally lowered its magnificent horned head into a pool of water and drank to quench its insatiable thirst as infection set in, rendering it defenseless (but still conscious) as the Komodo's feast began.

Here was an animal programmed by evolution to assemble all the tools nature provided for its survival — not only power, quickness, vision, and smell, but also cunning and patience. This is why the late evolutionary biologist Paul MacLean considered Komodos to be kindred spirits. The reptilian brain is alive and well in humans. We have as much in common with a near-extinct lizard as we have in common with our pets.

How dogs and reptiles differ is evident in their family arrangements. Male and female Komodos often mate for life, yet this too is a territorial imperative. The couple share the onerous task of guarding the perimeters of their adjoining kingdoms. They do not, MacLean wryly noted, share the same bedroom. And they would just as soon have their babies for lunch as care for them.

If I were a spiritual person, I'd say Mel and I are soul mates. Would I say that about the climbing hydrangea that was my pride and joy for twelve years until it blew over in a storm? There'd been a death in my family; the hydrangea was express-ing its sympathy, a friend suggested. Being a gardener, I was tempted to agree, knowing that some plants do look out for each other.

In plants the survival instinct is all about protecting the spe-cies. An *individual* survival mechanism is not built in. Straw-berry, clover, reed, and ground elder put out long underground runners that form networks—information highways, if you will. A Dutch botanist found that if one of the network plants is attacked by caterpillars, the other members receive a signal telling them to make their leaves hard to chew on and less de-sirable to hungry caterpillars. Word spreads fast. Bamboo and many invasive grasses use the same system. The network's only drawback is analogous to what happens when a hacker invades a computer's operating system. Viruses may enter the plant via the leaves, find their way into the stems, and be passively trans-ported to all the network members, where they cause new in-fections.

What interested me about this research wasn't that plants communicated with one another but what their conversational style said about where they fit in the evolutionary scheme of things. Smell marks the beginning of brain development, the seminal moment that sent plants in one direction and moving (and eventually thinking) creatures in another.

The tiny sea squirt has a brain and notochord (a sort of backbone that is a precursor to the nervous system) only while it's young and swimming in the ocean. The creature ditches this pintsize organ once it anchors itself to a rock. No movement, no brain. Olfaction may have been a navigational tool unique to species that were built to explore and migrate. Among animal species some patterns are clear: Komodo dragons are antisocial creatures who interact (fight) with their own only to protect their turf, which represents the most basic ingredient of survival: food. Rats, dogs, birds, and bees run in packs, and to them personal safety depends on the group. That means family, so their chatter centers on mating and reproduction. Humans are a blend of those instincts, but our species' talk is mediated by a strong sense of self. Language allows people to speak with one another about our place in nature and the universe, and about paradox—good and evil, the sense of being alone, like the lizard, but also connected by intelligence and curiosity to every other living thing. We talk about that.

Scientists are beginning to understand the neural substrates humans share with other species, from songbirds to dolphins to our closest relatives, the primates. Neuroscientist Joseph LeDoux believes that even in humans, language is instinctive. Forming a sentence is an unconscious act; imagine how long it would take to get the sentence out otherwise. You don't start from scratch and put words together according to the rules of grammar every time you speak. Talking and marking aren't really so different, are they? This suggests that language, which is only slightly better understood than smell, is not the slave of the thinking brain, as is usually assumed, but is tethered to emotion. When you use language to communicate your feelings, the neocortex has limited authority. It can't tell you that your feelings are silly or wrong. Its role is to help you get in touch with feelings and then get them out by tapping into learned responses stored in the limbic system. "Consciousness is important," LeDoux wrote in *Synaptic Self*, "but so are the underlying cognitive, emotional, and motivational processes that work unconsciously."

So while language isn't intelligence, it's probably what made us smart. It's what makes all species smart. The world's first aroma cues sent urgent messages regarding, for example, the presence of a disgusting (read: maybe toxic) food. Eventually the repertoire grew. The more humans evolved as a social species, the less they relied on physical prowess to survive. Strategic collaboration required a means of communicating that was complicated, quick, and neuron-intensive. Then *Homo sapiens* needed a bigger brain so communication could continue to develop, but the skull couldn't hold a bigger brain. Something had to go.

Olfaction was one obvious place to make cuts. Ever since primates began standing upright—bipedalism improved visual awareness—smelling had been compromised by the simple fact that heavy odor molecules never rise to nose level. Dogs still sniff to say hello and mark territory, but in primates, visual communication cues grew more subtle and nuanced. In humans, these cues (a curled lip to register a foul odor is still part of our human DNA) developed into spoken words. The curled lip meant "rotten," which also came to mean "faulty," as in a "rotten" idea. Out of language evolved abstract thinking. Evolution didn't put words and smell totally at odds. Studies of nasal cycles in humans indicate that a constant strong flow of odors to the left hemisphere, the language side, of the brain makes a person more articulate. Maybe that's because odors arouse the memories behind the words.

Still, a red balloon isn't merely red; it has a unique verbal descriptor. A tune is *melodic*, a whisper *low*, and a gunshot *loud*. Sugar is *sweet* and lemons *sour*. Wool *scratchy*, velvet *soft*. Language may not be the slave of the thinking brain, but it has risen in importance in lockstep with it, and pretty much in tandem with smell's decline in the gene pool. That is nowhere more evident than in what Andrei Codrescu bemoaned as our sparse smell vocabulary. Richard Axel's primal sense has to borrow from taste — words like *sweet* and *sour*, *fungal* and *fume-like* — or resort to analogies. There is no word in any language specifically meant to describe the smell of an old sock.

17

THE PAINTED SMELL

I'VE ALWAYS BEEN a poor sleeper. Even when I had small children and was exhausted most of the time, I felt lucky if I got four hours of shuteye a night. If some household project didn't occupy my restless mind during those sleepless predawn hours, I read a book.

A friend had given me *The Kite Runner* for Christmas. Months later I had a go at it and after a week I'd fought my way to page 141, where the book took hold of me. For the first time in months the words were not jumping around but invisible, and I was no longer reading in my soft, warm bed but struggling to breathe in the black bowels of an oil truck. I must keep still, and quiet. My father's hand is touching my sleeve. The glowing white dial of his wristwatch emerges out of the darkness. He whispers to me to keep my eyes on the glowing dial. The smells in the truck that had seemed suffocating vanish.

The main character, a young Afghan boy named Amir, keeps his eyes glued to the wristwatch dial for twelve hours, until the oil truck has crossed the border to freedom and escaped from the Soviet invaders. Even though the boy knows the fumes in the tank are harmless and will soon fade from his awareness the way other smells, like the smell of garlic in his mother's kitchen, do, he cannot bear the stench until a visual distraction anchors him once again to a familiar here and now. His eyes—vision—save him. Amir survives twelve hours in the black belly of an oil truck by staring at his father's watch.

My heart raced and then I too was gasping for air. The fumes *were* suffocating—and it never occurred to me as I read that I wouldn't have been able to smell them.

This was progress. The memory of smell was better than nothing, and books restored that memory, at least in the abstract. The primary sense can't conjure up smells in the brain the way vision can summon pictures, and hearing can recall sounds. But memories can be triggered by a familiar scent. Writers love this sense's mysterious talent as much for its literary usefulness as for its mystery. There is no solid explanation for what is often called the Proust phenomenon.

I felt oddly elated, as if my mind had been a caged bird set loose in a room; I too had been set free, if only for a few minutes. Reading was still hard work. I told my husband it was even harder than working out at the gym, something else I was making myself do because I remembered the effects of endorphins and trusted that one day those endorphins would wake up. Biology would force them to, just as biology in the form of a pill had forced the foul smells out of my brain, leaving me entirely anosmic. I knew, or rather believed, that without this torturous *trying*, the slow stitching together of mind and body, emotion and thought could not happen. It was as necessary as changing the dressings on a painful wound, or climbing back on a horse that keeps tossing you into the dirt, or picking up a musical instrument after you've botched a recital, or learning to like raw fish when you've grown up on meat and potatoes and always cast a wary eye at even the cooked seafood entrées (who would

order that?) on steak-house menus. Hard, hard, hard. But necessary. *Put one foot in front of the other.*

I went to the library and brought home a pile of books I'd
been meaning to read, hoping something in them would grab
me the way that passage in *The Kite Runner* had. Deprived of
storytelling's magic, I became obsessed with exposing its mechanical underpinnings. This was better than staring at a fuzzy
and unintelligible page, and something I could get my discom-
bobulated brain to focus on. I felt like a college student parsing lines of poetry. Caroline hated the way writing papers and
"overanalyzing," as she put it, stole a book's pleasure from her.
She was always too worried about getting a book "right" to get
lost in it.

I noticed an abundance of olfactory references in many of
the best of the books I read. And I began to commit that sin
of analysis purposely. I could see patterns in the authors' use of
the primary sense to convey primary things, to say the unspeak-
able and illuminate the imponderable. I noticed that smell was
particularly useful in first-person accounts when the author
wanted the reader to view the world through the lens of an
honest narrator, someone plainspoken and direct, someone either too young or too unworldly to obscure his observations in
self-conscious rhetorical flourishes.

Smell makes an excellent character witness, testifying to the
narrator's innocence and thus candor. Some of the opening
lines in *East of Eden,* one of Caroline's favorite books (unsullied
by modern English lit class), are an obvious example. The narrator is a farmer, a good man of simple background. How do we
know this? Steinbeck transplants his own boyhood memories
of California's Salinas Valley into the head (and through the
voice) of his narrator, who sees things clearly through his nose.
He isn't a thinker but a doer. Steinbeck knew that if he could

get readers to smell that valley, they'd end up inventing their own valleys and making the story theirs too. It would seem real and alive, personal, even if its description was minimal.

Since smells can't be conjured in adjectives, Steinbeck began his novel with a carefully worded inventory of fragrant things: "I remember my childhood names for grasses and secret flowers. I remember where a toad may live and what time the birds awaken in the summer—and what trees and seasons smelled like—how people looked and walked and smelled even. The memory of odors is very rich." The narrator trusts his nose, and this makes him trustworthy.

Marilynne Robinson's novel *Housekeeping* opens the same way. The narrator, Ruth, has had a hard life. The book is an extended backward glance at that life, and it's heavily infused with odors. It starts with the smell of a mysterious and menacing presence, a lake that takes the life of the narrator's grandmother's husband and the rest of those on board a train that had sunk "like a stone" or slid "like an eel" off a bridge into the lake and disappeared with scarcely a trace. We learn that the wives of two victims left town at once, driven away in their sorrow by the lake's inescapable watery smell. The word *they* is another device that lends credence to the event and its narrator, who was not yet born and so was not there that horrible night the train derailed. She relies on others' accounts. But were *they* reliable? Possibly. *They* said that the two women who left town "could no longer live by the lake" because "the wind smelled of it," as did their drinking water.

The lake is not made sinister in the usual way. Its smell is not fishy or fecal. Its smell is "watery." It smells like nothing except what it is made of. Yet it is unearthly, this smell. The odor's compelling peculiarity makes the lake's presence as a character all the more ominous. Ruth's grandmother is the only one of the three newly bereaved wives to stay in town—she has three daughters to raise—and in a beautiful paragraph she is described in her widow's black "performing the rituals of the ordinary as an act of faith." She hangs laundry and fights a strong wind off the lake. "It smelled sweetly of snow, and rankly of melting snow, and it called to mind the small, scarce, stemmy flowers that she and Edmund would walk half a day to pick, though in another day they would all be wilted."

I trusted Ruth's imagined version of this marriage because I trusted her emotional memory. For one thing, I hadn't ever read a more honest account of what it's like to go on a wildflower hunt in early spring while knowing the hunt is futile. Pried loose from the gravelly dirt, alpine plants seem to wilt more from homesickness than lack of care. In fact, in most cases, it's too much care that kills them. Edmund lifted the flowers "earth and all" into a bucket with a trowel. They died because they "were rare things, and grew out of ants' nests and bear dung and the flesh of perished animals."

As the pair trudged home with their doomed treasures, "the wind would be sour with stale snow and death and pine pitch and wildflowers." Death? Though it was early spring, the sour smell was that of the waning winter. Last year's dead leaves and flowers, now ripe with mold. Soon would come the "resurrection of the ordinary," and in spite of the wind Ruth's grandmother looked forward to when "the skunk cabbage would come up, and the cidery smell would rise in the orchard, and the girls would wash and starch and iron their cotton dresses." The watery smell is only faintly suggestive of death, and its intimation of those unrecovered corpses is more chilling and sinister than any real corpse thrown up onto the shore, rotting and bloated, would have been.

The violence and inconclusiveness of the derailment is embodied by the old lake, "smothered and nameless and altogether black." Especially in spring, when the water of the lake is "suspended in sunlight, sharp as the breath of an animal."

A neighbor and long acquaintance of mine, the author Patricia Hampl, was writing a memoir about her parents, *The Florist's Daughter*. Her mother, a feisty Irish lady and devout Catholic to the end, had recently passed away at eighty-five. Hampl's mother had been born the same year as mine, 1917, and, also like my mother, had been living with the damage caused by a stroke. Even the dream life Hampl describes her mother creating for herself toward the end was like my mother's, romantic fantasies involving various male composites. *The Florist's Daughter* is a tribute to things universally important and dear, and doubly so to someone who happened to be born and raised in what Hampl and I both still refer to as "old St. Paul."

Old St. Paul was made up of Irish and German Catholics and just enough WASPs to give it a sense of entitlement and a level of class that was presumed lacking in its more openly ambitious sister city of Minneapolis, which was full of Swedes and Norwegians too busy to care. They left old St. Paul in the dust, with only our old neighborhoods to brag about, our Victorian homes preserved not by foresighted urban planning but by the town's lack of commercial success. Patricia Hampl's dad hired me when I was a teenager to work as a designer for the business he managed, a carriage trade florist called Holm and Olson. People in old St. Paul still used expressions like *carriage trade* in the sixties, even though the carriages had long since vanished. I was a junior in high school with no experience designing anything. Nevertheless, I believed him when he said he thought I showed promise. I learned the truth in Hampl's book—her dad did favors for women like my grandmother, a good customer.

Patricia worked for her dad too. Her coworkers called her the chatterbox. Unfortunately, when I worked for Stan Hampl as a floral designer I was struck deaf and mute by embarrassment at my own shocking ineptitude. For eight hours a day I poked garnet roses into Styrofoam amid the tumult of deadlines and personal dramas being played out around me, trying to summon the courage to say something, if only an apology for existing.

My refuge was the greenhouse. Hampl remembers it as being like a farm, smelling of renewal. She "was willing to be enchanted" by the greenhouses and the silent, unglamorous (compared with the designers) men who cared for the tender plants. So was I. But the florist's daughter was accepted into the immigrant Eastern European fold as family, while I was an outsider and felt keenly my lower status, the girl brought on as a favor whose grandmother was one of what Hampl describes as "the ladies of the little curving streets of leafy Crocus Hill" who all trusted her dad "to follow them in their decorative plots . . . to supply the palette of their spring gardens, planted every fall, and then furiously dug up and discarded to be replaced with annuals every summer for a few short months until the whole business had to be rooted out and the process started again."

But these feelings aren't what present themselves to me unbidden whenever I enter a greenhouse and am engulfed by that inimitable greenhouse smell, as universally familiar as the smell of a bakery. Instead I feel a sense of relief triggered by memories of when I would escape the design room and creep into the greenhouse and sit down with a book on the damp cement floor, surrounded by camellias and azaleas and geraniums awaiting permanent winter homes in townspeople's living rooms. I was and am at home with the smells I love best: dirt, manure, flowers.

My muscles relax, even the ones in my forehead that had begun digging a permanent furrow long before I found myself at Holm and Olson, desperate to truly earn my paycheck. The florist's daughter's words remind me of other smells that take me back to that greenhouse. And to the design room too. When the powerfully fragrant Casablanca lilies in my garden are blooming, I'm packing Easter lilies into "the company's narrow, powder-pink boxes," carefully wrapping them in the "waxy green tissue to be sent on their way in the forest green Holm and Olson trucks," as Hampl recalls the ritual.

Never lilacs. They were too fragile. But May still means lilacs in old St. Paul.

> And here it was—lilac time. And lilacs my favorite of all flowers. Actually, a florist father would seem to work against a passion for lilacs. For lilacs, being practically wild, abundant, and free, are dismal failures as a retail commodity. But at this time of year, St. Paul, finally unfrozen, is always a lilac-town. The lilac is our consolation prize, a post-winter badge of honor. The scent lies heavy in the mid-May air, . . . the incense of ages . . . In mid-May lilac even overpowers the rank garbage cans along the narrow alleys of the Crocus Hill neighborhood where, it seems, the oldest, most profuse lilac canes flourish.

Spring was coming and with it the lilacs. Lilac time. The word alone—*lilacs*—brought visions of springs past. But no fragrance. I was again reminded that smell differs from hearing and sight in that it cannot be remembered. You can dredge up a tune or picture a long-lost loved one's face, but you can't conjure a scent out of whole cloth, even with the help of a memory. I would never smell spring again, even in my mind's eye.

18

SMELL, MEMORY

IN *Speak, Memory,* Vladimir Nabokov wrote that "nothing is sweeter or stranger than to ponder those first thrills [of childhood]." He remembered how he chewed a corner of a bed sheet "until it was thoroughly soaked" and then wrapped a candy Easter egg in it tightly "so as to admire and re-lick the warm, ruddy glitter of the snugly enveloped facets that came seeping through." Such recollections are unimpeachable, he concluded, because they "possess a naturally plastic form in one's memory which can be set down with hardly any effort." Memories are more elusive if they lack corroboration by all the senses, especially the sense of smell. Without it, "I have to go by comets and eclipses, as historians do when they tackle the fragments of a saga."

Moments lost in a book were still rare as winter wore on. And it would be a long while before a book could put me to sleep. Sometimes I would give up on reading. One night after Cam had fallen asleep I made my way like a sleepwalker, or like one of those ghoulish zombies in a horror film, to my attic office. I turned on the computer and opened a folder on my desktop where I kept pictures of my garden. Minnesota winters are so long, you forget what summer looks like.

The first photo was of my back porch. I'd trained the fragrant American Beauty rose to a trellis on one of the posts, along with several other aromatic vines. The idea was to create an olfactory symphony—composed of notes, just the way perfumers did it—with roses and wisteria as the high and middle floral notes and herbs as the less feminine but zestier and longer-lasting low notes. The potted herbs sat in three rows on the porch steps, like a family having their picture taken. There was the prim mother (a scented geranium on a standard), the paunchy dad (sweet basil in a round ceramic planter), the gaggle of unruly kids (mint, mint, and more mint), the gnarled grandparents (sage and rosemary). There were various aunts and uncles too: lemon oregano, French tarragon, English lavender, and thyme. I had been thinking (then) of sending this out as a Christmas card.

While reading fiction remained difficult, I continued to devour science papers. It was comforting to read that science was trying to cure emotional problems caused by life-changing trauma. Any scientifically proven connection between smell and mood got my attention. I'd read somewhere that sniffing helped lift subconscious smells locked in memory to the level of consciousness. Did one need a working nose for that?

Scientists like Richard Axel call smell the primal sense because of its primary importance to survival in ancient species. In conjunction with the fear response (fight or flight) it was the first line of defense against predators. This role seems to have made the oldest parts of the brain the most stable and resilient; the olfactory cells are the only neurons that repair themselves, and

smell is the most unerring of the senses when it comes to at-
taching an emotion (first and foremost is fear) to an event. This
attachment is so strong that the original event is the one most
likely to rise from the depths of submerged long-term memory
to conscious awareness on the strength of a whiff of an odor.
No subsequent event associated with the smell is as quick to
present itself as the first one. The emotional power of the event,
whether positive or negative, also influences smell's effective-
ness as a trigger. First the nose evaluates whether a smell is a
threat, based on experience, and then sends an action message
(such as fight or flight) to the rest of the limbic system; only
then does the mind register the odor's name. In all these ways
smell differs from the newer senses of sight and hearing. Rachel Herz thinks humans would not experience emotion at all
if it weren't for smell.

Trygg Engen explained in his *Odor Sensation and Memory*
that smell has no identifiable attributes of its own but exists as
an inherent part of what he calls "a unitary, holistic perceptual
event. It is as if the memory of an odor is protected somehow,
so that other experiences don't interfere with the memory of it,
whereas pictures and sounds don't have that same protective effect." The feelings an odor triggers can change, however. For
months after a house fire you might experience intense fear and
anguish when exposed to the smell of smoke. As time goes on,
the smell will revert to its default position—the pleasure of
those long-ago evenings at summer camp when you roasted
marshmallows around an open fire. The nose never forgets anything.

Researchers at the State University of New York have shown
in animal studies that the memory center in the brain is like
a reusable storage disk, thanks to an enzyme called protein
kinase M zeta, which preserves long-term memories through

persistent strengthening of synaptic connections between neurons. By manipulating kinase M zeta, the researchers believe they can effectively wipe out negative memories in human subjects without putting other data or the storage device itself (the hippocampus) at risk. Bad memories encompass a broad landscape, from the terrifying images (smells and sounds as well as sights) that haunt people with posttraumatic stress disorder to phantom sensations. Alzheimer's patients may also benefit from the discovery of the enzyme, which is bound up in the tangles that destroy existing memory and block memory storage.

I fantasized about teaching myself a whole new way to call up my stored memory of plants through using touch. Plants have exquisite textures. Some are like velvet and others prickly; some feel almost liquid to the touch and some are as smooth as glass. Leaves can be as soft as a baby's cheek, furry, or bristly. There are even plants that feel like sandpaper. Ruminating on the mysteries of smell science, I pondered fresh quandaries, such as why touch sensations, so like smells in that they run the gamut from pleasant to painful, lack the emotional component that makes smelling a rose so romantic and smelling a dead animal so rank. Why did evolution decide not to route touch through the limbic system? Touch lacks smell's influence over whom a person falls in love with and what risky situations to avoid. What determined that the nose should be given power and influence that even the eyes and ears do not possess?

We'll probably never know. Whether olfaction's special ability—the power to heal its own neurons (called neurogenesis)—might exist in the rest of the brain remains unclear. But we do know that the brain is plastic and rewires itself in response to the demands put on it. People who are blind develop a more discerning auditory system. Pianists train their fingers to perform feats of physical coordination that far exceed the innate potential of even an Oscar Peterson. I decided that I would retrain my insensitive, clueless, brain-dead fingers. I would teach them to smell.

That night in my attic office, holding that thought, I turned off the computer and entered the still, dark hallway. I didn't have my nose to guide me as it used to when I stayed up late working, making the darkness familiar — my smelly old house. Now I consciously ran my open palm along the rough plaster walls, over the cool, smooth banisters. So familiar and so novel at the same time. Before brushing my teeth I let the tap water caress my wrists and tried to deconstruct *wet* as if it were . . . what? I was reminded once again that there are no words that specifically describe odors, only allusions to other things. A rotten-egg smell. The smell of sour lemons. Whereas objects painful to the touch are described as *scalding, sharp, jagged.*

The next morning I tried to explain to Cam this new project I'd come up with, but, like so many predawn epiphanies, it dissolved the instant I put words to it. My smell-touch idea couldn't stand the light of day, that much was clear. Touch is not smell. It cannot be rewired to become smell. Touch is touch. I went to the refrigerator to get out the coffee beans and then noticed that apparently the dregs of yesterday's coffee were still in the pot. I mumbled an apology. Cam pulled a mug out of a cupboard and emptied the contents of the coffeepot into it.

"I made it twenty minutes ago," he said, smiling. "Guess you can't smell it."

He was making progress, anyway. He would be over this in no time. He'd accept anosmia for what it was, stop fighting the diagnosis, and move on. Easy for him. For me the casualty numbers kept climbing. Over tasteless coffee and slimy cereal I told Cam that I couldn't seem to write. Gardening was an impossible topic. Reading didn't take me out of myself, not really, unless I was reading a printout of an article that had appeared in *Cell, Science, Nature Neuroscience, Neuron, Chemical Senses,* or some other science journal. Or an e-mail from someone whose name was attached to one of those articles and who had been kind enough to answer a question from a nonscientist.

As the weeks and months went by and my nose did not improve, I began to count my blessings. For real. Dr. Cushing had been correct when he told me that being a writer without smell was better by far than being a chef without it. I was finding that I could forget all about my nose for minutes, hours, even days by immersing myself in smell science.

But I was not my old self. I'd lost my memory not of how the past looked, which is mutable and untrustworthy, but of how it smelled, which is inviolate.

I still wanted to believe that a clinical approach to understanding olfaction would help me stop obsessing on catastrophic (outlandish) consequences — such as this new one I was chewing on: *You have only so much time to gather up all your memories and write them down, capture them somewhere (in the computer?) before your brain presses the Delete key.* If you think that's crazy, I said to my husband, how do you account for Proust? How did his brain suddenly give him back his lost childhood after he tasted a madeleine soaked in tea? Why couldn't that happen in reverse? I was worried I'd forget what people, places, and things looked (and sounded and felt) like if I didn't have smell to pack things together for storage in the hippocampus. Surely they'd become disconnected, scattered, and inscrutable.

Northwestern University neuroscientist and smell researcher Jay Gottfried described smell as a paradox in that it's both speedy and sluggish. Vision is speedy; a person can recognize a face in one-tenth of a second because the information from sight goes straight to the thalamus and high brain. Smells end up in the orbitofrontal cortex, the same place in the high brain that allows humans to recognize faces; it just takes an odor longer to get there. When this region is damaged, both face and odor recognition are lost. Professional wine tasters in France demonstrated the lag time between sight and smell when they were fooled into mistaking dyed-red white wines for red burgundies. Imagine their embarrassment when told *mais non!* They'd been tricked by a timing issue. The visual was signaling *red* in the thinking brain while the white-wine smells (lemon, straw, melon, toast, and grass) were still hung up in the limbic system.

Smells may be slow to register cognitively, but they operate with superb efficiency subliminally. Ever notice how swiftly your dog makes the dash to the door when he sniffs an unfamiliar dog outside? Compare that to his response to a passing car. He'll go to the window, but with nowhere near the same vase-toppling intensity. In humans too, smells carry potent emotional messages. This is partly why elderly people with dementia can still summon old memories with remarkable clarity even though more recent ones won't stick. An emotion-laden, long-ago fragrance can put an old man right back in the grove of cedars where he built forts as a kid; it can put an old woman into the '56 Chevy where she kissed a boy for the first time (his Old Spice and tobacco breath may be all that's left of him).

Then, too, smell-induced memory triggers weaken not with time but with overuse. The brain becomes desensitized by repetition. That's why sitting down to the fabulous pasta you learned to make on a recent trip to Italy doesn't have the emotional impact of pulling up a chair in front of a bowl of Chef

Boyardee. It's been decades since you last smelled that inimitable blend of plasticky sweet ketchup and oversalted meat. A whiff of Bolognese sauce may take you into orbit, but it won't take you back in time.

Don't imagine that those old memories are authentic, though. The brain does not photocopy the past. Memories are remade with each act of remembrance. So while the same smell triggers the memory, the remembered events of the past are revised. They are reinterpreted, if you will, by subsequent incidents and emotional responses. Why? Memory's evolutionary purpose was to enable learning; thus, it *must* be highly flexible and subjective.

Richard Axel concluded his speech to the Nobel committee with a nod to Proust's remarkable insight that only smell can make the past present. Scientists and literature scholars alike have been pondering "the Proust phenomenon" ever since *In Search of Lost Time* was published in 1928. Proust knew intuitively that "olfactory sensory maps must be plastic," Axel said, "with our genes creating only a substrate upon which experience can shape how we perceive the external world." The great French novelist wouldn't have chosen quite those words, but *In Search of Lost Time* makes the point again and again.

The phenomenon is more complex than it seems. Researchers who assumed they could resurrect old memories in subjects by exposing them to fifteen familiar smells were disappointed. An odor has to be intricately detailed and nearly one-of-a-kind to unlock a memory. Proust's happy childhood memories came on a whiff of a tea-soaked cookie. This wasn't just any cookie but a madeleine whose buttery-sweet fragrance had been released through chewing and swept up through the retronasal passage. There it was joined by the aroma of, again, not just any tea but the one already present (having infused the cookie's taste), now rushing its smell of lime-scented linden flowers to the brain by way of the nose.

So lovely, and so familiar. Proust at first couldn't remember the last time he'd smelled this scent. Then it came to him—yes, it was in Combray, the French village of his lost childhood. As a boy he'd made frequent visits to his invalid aunt who lived upstairs in his parents' house and who always treated him to tea and a cookie. Decades later, when Proust's mother offered him the same tea and cookie (in part to provide her son with a brief respite from a melancholic state he described as the deadening effect of dull habit), his forgotten childhood returned. The tea and cookie released old memories as if they'd been captured in a frozen waterfall and then freed by a spring thaw. A wave of euphoria came over him, and then the memories. Torrents of words followed. Proust filled thick volumes with the memories that his adult mind, for reasons both mysterious and practical, magically set free.

The strongest odor-induced memories tend to be of moments of delight. These memories are not quite the same neurologically as memories that seem indelible because one was intensely "present" and alert at the time of the event; memories that haunt war veterans fall into that category. Both the nostalgic and the intense types of memories may be triggered by an odor, but the sudden recollection of traumatic battlefield memories can be problematic. Marcel despised the smell of varnish because it took him back emotionally to those nights when he was a small boy and his mother sent him upstairs to bed without a good-night kiss. That detestable wooden staircase, icon of his banishment, was a potent trigger for separation anxiety all his life—not the staircase itself, but its smell. And yet, though

Proust hints at a troubled childhood, his most strikingly detailed odor-induced snapshots are of good times. It's as if the brain intentionally filters out emotional triggers that might add up to an accumulated burden of sadness too heavy to bear.

On the strength of a sniff—okay, a few sniffs—Proust emerged from a long and incapacitating struggle with depression and what would now be called obsessive-compulsive disorder (even those who know nothing else about Proust are usually aware that he lived most of his adult life in a cork-lined room). His mind finally focused on something outside his immediate circumstances, this familiar smell, this all-powerful joy. After several false starts (and as the tea-soaked cookie lost its power), the memories took shape even without the odor, restoring his hope and happiness by rekindling his imagination. The creativity that had been so unselfconscious in childhood brought his "true self" back to him. It endured until the end of his writing life.

There are skeptics, of course. Some dismiss Proust's tale of remembrance as pure fabrication, arguing that he'd been casting about for a fashionable literary topic and seized on Freud's theory of the unconscious mind. Non-navel-gazing storytellers, such as Steinbeck, Flaubert, and Hemingway, who render smells in gorgeous prose, are olfaction's real champions. Science writer Jonah Lehrer disagrees. He thinks Proust may indeed have been influenced by the work of Freud, but he was also influenced by hard science. Up until 1900, scientists thought neurons were linked together, forming a sort of meshlike net. Then the flamboyant Spanish scientist Santiago Ramón y Cajal "stared at thin slices of brain under a microscope and let his imagination run wild," wrote Lehrer in *Proust Was a Neuroscientist*. In a classic example of top-down science, Cajal guessed that the spaces between neurons, called synapses, were the communication centers for electrical signals throughout the brain. He was right. Did Proust, hearing of this, let his imagination run wild, as Cajal's had? Subtle shifts in synapse strength ease the communication process that allows neurons to trigger memories, Lehrer explained. "The end result is that when Proust tastes a madeleine, the neurons downstream of the cookie's taste, the ones that code for Combray and Aunt Leonie, light up. The cells have become inextricably entwined; a memory has been made."

Memory research has until recently used a scientific model based on enzymes and genes; mechanisms could be studied only through a process of training lab animals, "bullying" their neurons into altering their synaptic connections, as Lehrer described it. "Senseless repetition seemed to be the secret of memory." But while such experiments do show how the brain adapts, they don't "encapsulate the randomness and weirdness of the memory we live in," the way memories "change and float, sink and swell [and] disobey every logic" and defy our ability to "know what we will retain and what we will forget."

In 2003, Kausik Si, a researcher for whom Lehrer worked in Nobel laureate Eric Kandel's lab at Columbia University, published in *Cell* what Si hoped would be a breakthrough theory of memory. Si had earlier discovered that a certain protein called a prion can erase or create memory by changing shape. Prions are stimulated by serotonin and dopamine, two neurotransmitters produced by thinking. How could Si resist hypothesizing a link between such renegade particles and the kind of memories that "disobey logic"? Having been activated by the first taste of, say, a madeleine dipped in linden-scented tea, the protein marks a specific dendritic branch as a memory, Lehrer wrote.

The protein will patiently wait, quietly loitering in your synapses. One could never eat another madeleine, and Combray would still be there, lost in time. It is only when the cookie is dipped in the tea, when the memory is summoned to the shimmering surface, that [the protein] comes alive again. The taste of the cookie triggers a rush of new neurotransmitters to the neurons representing Combray, and, if a certain tipping point is reached, the activated [protein] infects its neighboring dendrites. From this cellular shudder, the memory is born.

Gordon Shepherd dug up an old article for me to read in which he outlined why Proust's successful efforts to evoke the smell of the tea-soaked cookie were in sync with what scientists have come to understand about the neurology of smell. Scholars had tended to regard this effortful conjuring as Proust's prerogative to describe the episode as he saw fit. He was an artist, not a scientist. The suspenseful telling only made the episode, and its essential insight, more powerfully thought-provoking. Now scientists know that Proust's recollection of his recollection was spot-on.

The narrator, Marcel, freely admits that intellect did the heavy lifting after smell waltzed by. When he seems to later disqualify intellect's power, Proust is really just asking his readers to focus on the power of memory itself, the way it galvanizes our artistic souls by bringing us back to childhood. By dramatizing the *difficulty* of relying on intellect to stimulate memory, he shows us the breach, created by the thinking brain, between present and past, between our true selves and the strangers we so often become.

Part III

TOXIC SMELL

19

A BARRIER BREACHED

ABSENCE OF SMELL was better than phantosmia after all. A brain filled with vile odors can't keep track of the car keys, let alone ask reasonable questions. Questions raised by phantosmia, for instance. The questions were becoming more persistent, like early-morning dreams, the most vivid and usually horrific, that you can't shake off. Why did you keep falling down when the monsters were chasing you? Why did they all look like your boss?

The question nagging at me now was this: Why does the brain *care* so much about smell? It doesn't go nuts when you lose a toenail. Well, for one thing, the toenail will grow back on its own. Toenail tissue, like all human tissue except the kind in the brain, quickly regenerates new cells. Brain cells are iffier. New research suggests that some brain cells may repair themselves some of the time, and although this may one day allow science to manipulate genes and harness neurons' ability to divide and multiply, for the moment people with paralyzing spinal cord injuries are still confined to wheelchairs. The damaged neurons, no matter how threatening their loss is to the individual's survival, do not automatically begin regenerating.

That cells in the olfactory system do have the capacity to heal suggests — even more powerfully than phantosmia does — that the brain cares a lot about smell. We think of the nose as a highway for odor molecules, but it's actually a filter. That's why a dog's snout is huge compared to a human's nose. The most hazardous airborne molecules are big and heavy, so they hover just above the ground. A dog's big snout is designed to filter them out before they can reach the brain and lungs. Gordon Shepherd thinks the human nose shrank in size as our ancestors took to walking on two legs instead of four. We don't need big filters to protect us from lightweight airborne volatiles.

Or do we? As mankind spews more and more toxic and extremely tiny particulates into the air, our nasal filters may not be up to the task of keeping our brains pollution-free. There's no telling how our brains — our genes, actually — are responding to this new type of outside threat. It hasn't escaped the attention of brain scientists that the olfactory system is the only brain function able to repair itself, and this has enormous implications for neuroscience as a whole. For one thing, it suggests that Mother Nature wasn't willing to risk exposing the brain to irreparably damaging substances capable of rising to human nose level. But not all toxic odorants are big and heavy. And people aren't always high above them. What about infants? They spend a lot of time rolling around on the ground. A young and developing brain is more vulnerable than an adult brain. Scientists now think smell cells regenerate because of such factors — and because they're being destroyed more frequently than we realize.

Moreover, while some people regard anosmia as an insignificant problem, the human limbic system may not share that bias. Its design predates *Homo sapiens*, and as far as the limbic system is concerned, a loss of olfactory function is as hazardous to a human as it is to any animal—and even more hazardous than a spinal cord injury. There is no doubt that our oldest postprimate ancestors—the first humans—depended on smell much more than we do. Even while the genes for smell were adapting to new tasks, these people relied on olfaction to sniff out both dinner and danger. They probably practiced smelling, though not consciously, just as professional noses develop their superior sense of smell through years and years of practice.

It used to be assumed that the blood-brain barrier protected brain cells from toxins that circulated through the rest of the body. The blood-brain barrier is a physiological mechanism that prevents certain substances from entering the brain, minimizing toxic exposure. But brain cells can be negatively affected by substances inhaled through the nose, so all manner of chemicals are now under scrutiny, from secondhand smoke to car exhaust to nano-size particles used to make computers lightning fast (among other applications). Scientists at the University of Rochester Medical Center showed that when rats breathed in such tiny materials, the chemicals made a beeline from the nasal cavity to several regions of the brain. Factory welders routinely inhale ultrafine manganese oxide as well as bundled "rods" of zinc, carbon, and gold particulates used in the miniaturization of electronic, optical, and medical devices. Many consumer products, such as toothpaste, lotions, and some sunscreens, also contain such particles. The University of Rochester study is part of a five-year, five-and-a-half-million-dollar investigation of the effects of nano particles on human cells; the research was launched in 2004 by the Department of Defense.

A study of children in California found that black mold, a malignant fungus that thrives in damp basements (and all over

New Orleans since Hurricane Katrina), causes memory problems when it's sniffed. Researchers at Michigan State University had mice sniff minute amounts of black mold. Each mouse suffered a significant loss of smell, as well as inflammation that spread into the higher brain's memory circuits. In some, the olfactory bulb was destroyed. Newer studies report similar links between air pollutants and brain disease. Dogs who died in polluted Mexico City were found to have brains laced with lesions identical to those that cause the symptoms of Alzheimer's, while dogs exposed to rural air only were lesion-free.

Human subjects who volunteered to breathe in diesel fumes for several hours in a lab subsequently had abnormal EEGs. This diesel study points out another interesting proclivity of the smell brain: Sniffers of the toxin all agreed that the fumes smelled bad. Some said they made them feel nauseated. Sniff at your peril, our noses seem to be saying. But one person's bad odor may be another's olfactory delight. For a pair of newlyweds painting their first apartment, the happy experience may even turn a dangerous chemical — the VOCs in household paint — into an aphrodisiac.

Nevertheless, most potentially toxic cleaning agents and home-improvement products, such as bleach, ammonia, solvents, paint strippers, glues, and insulating products, are offensive to the nose. Their naturally off-putting odors have to be masked by synthetic copies of odorants such as pine, lavender, or linen. (Benjamin Moore, the paint brand, has held its own in a tough economy thanks in part to expensive new products that are low in VOCs and marketed as scentless.) What's more, as those marketing types at Procter & Gamble and Clorox know, bad smells tend to linger a long time in the air. Why doesn't the brain adapt to the bad smells — in effect, tune them out — as quickly as it does most smells? Maybe for the same reason the substances smell bad. Whether its response is innate or trained, the olfactory system is saying that danger is present.

Congenital psychiatric and perception disorders are also being linked to olfaction. It's possible that olfactory environmental factors are involved in utero, and even earlier than that.

Patients with Korsakoff's syndrome, a horrific disorder precipitated by thiamine deficiency, suffer from anterograde amnesia, an inability to create new memories; the syndrome may be the result of gaps in the normal storage process. These gaps occur when the shortage of thiamine creates a chemical imbalance that essentially bores holes in parts of the brain tissue, especially the medial thalamus and its connections to the hippocampus and cerebral cortex, the pathway that mediates movement between the biological equivalent of the RAM and the hard drive (that is, the short-term and long-term memory).

These neuroanatomical black holes can have bizarre effects. Patients try to cover up their memory deficits by confabulating, making up information to fill in the gaps in their memories. One Korsakoff's patient was described in Oliver Sacks's "The Lost Mariner." This man had no short-term memory and had "misplaced" almost thirty years of his life. Another Korsakoff's victim inspired a film that starred the comic actor Peter Sellers; in it, the protagonist embellishes (beyond recognition) his former medical career. Interestingly, victims of Korsakoff's can't tell odors apart or recognize faces, and they often suffer debilitating depression.

Patients with Huntington's chorea follow a similar emotional and olfactory path. First they lose the ability to express visually (for example, with a curled lip) their revulsion for unpleasant odors. Soon after they lose the sense of smell. Psychiatric symptoms of this congenital disease, which eventually destroys the nervous system, include anxiety and depression, compulsive and addictive behaviors, and problems recognizing others' emotional responses.

The same year that Richard Axel took his Nobel home to Columbia, a team of neuroscientists and pathologists at Johns Hopkins raised hopes that Lou Gehrig's disease might one day be curable through gene therapy. The researchers had prolonged the lives of diseased mice by transplanting adult stem cells from the olfactory bulbs of healthy mice. The adult human olfactory bulb is now regarded as a potential nonembryonic source of stem cells to help those who've lost nerve cells due to injury or diseases like ALS and Parkinson's.

In 2007 a group of researchers in New Zealand happened upon what they called a cell "superhighway," a pathway that conducts stem cells from the olfactory bulb to the rest of the limbic system, apparently dropping some off via exit ramps throughout the brain. Their findings (inspired by a 1998 discovery of neurogenesis in rats and brought about by a decision to slice the brain in such a way that the pathway was immediately discernible) challenged the long-standing assumption that cells in the brain—other than olfactory cells—can't replace themselves. The lead New Zealand scientist believes that most new neurons are diverted before they reach the olfactory cortex, making them available for other brain repair.

In an interview on National Public Radio in February 2007, Richard Faull, who is an expert on brain diseases at the University of Auckland, described his team's extraordinary discovery. "It's like the freeway from Boston to Washington, D.C. It's actually got off-ramps going off to New York and all the rest of it. And we are seeing hints that cells are leaving this pathway well before the end of it."

Writer and former memory researcher Jonah Lehrer thinks the new cells are going to places involved in more important functions, like memory or motion. "These new brain cells may be vital to keeping these parts of the brain working." Story Landis, director of the National Institute of Neurological Disorders and Stroke, is hopeful that researchers can use the pathway to repair damage caused by the spectrum of neurological disorders, from Parkinson's and Alzheimer's to stroke.

I'd been without smell for fully five months. What had begun as a casual interest in smell was now an obsession. *Casual* is the wrong word. I was obsessed from day one — that being the day I started smelling things that weren't there, back in October. Internet browsing was being replaced by long phone conversations with important scientists. I'd even managed to arrange to visit the famous Axel lab at Columbia University.

I was greeted in the first-floor reception area by one of Axel's postdocs. Dan Stettler handed me a security badge, took me up in a tiny elevator to the tenth floor of the Julius and Armand Hammer Health Services Center, and ushered me into the lab, apologizing for the "awful stench." I didn't bother to remind him I couldn't smell it.

Dan admitted he was "a little OCD." He'd scrubbed down the entire lab the first week he'd worked there. It had been pretty disgusting, he said. Labs like this one wear their slovenliness like a badge of honor. Chemicals are smelly, especially the ones in a smell-science lab. In the postdoc's "office" (too glamorous a word for what consisted of a long counter with a computer on it), a graduate student was applying special dyes to the olfactory bulb of an anesthetized mouse. Dan bumped her arm and then apologized profusely. No harm done, she assured him.

Soon the sleeping mouse was ready for the trip to the two-photon microscope. Colors emanating from the mouse's olfactory bulb and olfactory cortex lit up as the mouse sniffed various fragrances: camphor, banana. (Humans, interestingly, don't experience odor sensation during sleep, but lower mammals do, presumably for the same reason that lower mammals smell more acutely than we do — their survival depends on being in a constant state of high alert against predators.)

The colors represent patterns of neuron activity, Dan said. He was in a good mood. Just the day before, after many false starts, he'd figured out how to display these same color arrays on a computer screen for his boss's PowerPoint presentations. These days, Axel relies on funding from various foundations to keep the lab supplied with postdocs, lab assistants, and two-photon microscopes; visuals are the bells and whistles for his presentations, but the "red meat," as scientists like to say, is the talk of medical breakthroughs.

Smell research is reaping unexpected rewards in the meat department. The study of all types of infectious diseases, neurological disorders, mental illnesses, and even cancers will certainly benefit from the work going on in this and other smell labs as scientists learn more about the mysterious ways in which environmental factors can trigger chain reactions that make us sick.

Yale researchers created the world's first olfactory map. They found a mutant fruit fly that had an olfactory neuron lacking an attached odor receptor; that neuron didn't respond to any smells. Using genetic engineering, the researchers created several mutant flies that each had a different odor receptor attached to the previously receptor-less neuron. They then tested each fly and established a receptor-to-neuron map. The scientists are hoping the map will serve as a model for the smell systems of insects that transmit disease, such as mosquitoes, as well as for smell systems of more complex organisms, including humans. Related studies with fruit flies published in 2007 showed how two genes allow flies to follow the scent of carbon dioxide; work on the molecular characteristics of odor perception in frogs may shed light on how malaria-spreading mosquitoes detect host odors.

Richard Axel frequently invokes the M-word. And why not? It's a jungle out there. Bill Gates isn't the first zillionaire philanthropist bent on eradicating disease. Centuries from now, the late Howard Hughes may be remembered not as an eccentric oil man who loved airplanes but as the man whose foundation, the Howard Hughes Medical Institute, funded the United States' top scientists. Richard Axel and Linda Buck, the biologist who shared the 2004 Nobel with him, both belong to this elite group. Howard Hughes investigators have access to more than money—they get to swap ideas with the best and brightest.

Not all are thrilled by the growing emphasis on "translational" research—where the focus is on tangible scientific consequences (breakthroughs, applications, products)—some want to get back to basic science. Caltech neurobiologist and HHMI investigator David J. Anderson, a former Axel postdoc, raised eyebrows when he abruptly shifted from mice to fruit flies in his quest to understand the connection between olfaction and behavior. Flies are an essential tool in biology labs because of their structural simplicity. Says Anderson, "In over four hundred million years of evolution, the same neural circuit organization has been conserved, even when the molecules themselves have not. And if this circuitry has been conserved, perhaps emotional responses have been as well." Anderson wants to know if, for example, the avoidance behavior of a fly to the smell of carbon dioxide represents an emotional response. This begs the question: What is emotion? And how did the earliest instinctive survival responses evolve into feelings?

Surely mice, being more like us than insects are, can provide solutions to medical problems more readily than flies can. To defend Anderson's work, some have suggested that a new type of antidepressant could come out of it. But to Anderson, that's a pretty big leap, and not his objective. The beauty of being an HHMI investigator, he says, is that the institute supports basic research. People like himself, Axel, Buck, and others interested in smell genetics can pursue any line of research that might lead to a better understanding of the brain. Switching from mice to flies in his lab required a huge learning curve for Anderson, one Axel also undertook when he brought fruit flies into his lab a few years back. "I feel like a graduate student again," Anderson says. He now thinks the structure of flies' wing positions may indicate their mood (for example, whether they want to avoid danger or mate).

Anderson and Axel have joined forces, applying their knowledge of fruit flies and mice to better understand how such behaviors evolved. Genes don't control behavior, but they design the neurons that do. By comparing similarities and differences between vertebrates and invertebrates, the scientists are hoping to figure out how and why their evolutionary paths diverged and what genetic mechanisms were at work.

Axel has found similar structural schemes in both mice and fruit flies. This indicates that all neurons that express a common olfactory receptor in these species have axons that converge on a single point in the brain.

Radically new treatments for a range of disorders may be on the horizon thanks to smell research. To find out about one new treatment, I didn't have to look much farther than my own backyard. University of Minnesota physician and smell researcher William Frey runs an Alzheimer's research center out of Regions Hospital in St. Paul. He and colleagues in Germany successfully implanted stem cells in the brains of rats and mice by a method they call snorting. The rodents are trained to sniff hard. Frey says one in ten cells makes it up the nasal cavity, through the skull, up the fluids in the olfactory pathways, all the way to the cerebral cortex, and finally out to the cerebellum. The farther the journey, the fewer the cells that show up in scans of the rodents' brains, but Frey believes he can improve the odds and that his technique will one day be the gold standard for stem cell implantation in the brain. Snorting is less likely than surgery to cause inflammation and other problems that could interfere with the cells' success in restoring brain function to damaged regions. Snorting is easier on the stem cells themselves and on the patients. And, unlike rats, people don't have to be trained to inhale sharply.

Northwestern's Jay Gottfried has nothing against smell research but deplores the lack of attention paid to smell itself. He complains that olfaction gets little attention in neurology textbooks and that this is a glaring omission. Why? Because smell dysfunction is a valuable diagnostic tool.

In other words, smell-dysfunction research itself is of minimal consequence unless it sheds light on how a toxin disturbs the olfactory system on its way to more "important" regions. Collateral damage. As I mined the medical literature for links between smell dysfunction and other diseases, it was all too obvious where priorities lay. Anosmia and phantosmia were side effects, symptoms, clues. They were not considered serious illnesses.

I seemed to have come full circle, for the truly serious illnesses were the very same disorders I'd imagined I had in the fall—Parkinson's, MS, schizophrenia, Alzheimer's. I could now toss in Lou Gehrig's disease, Huntington's disease, and Korsakoff's syndrome. With so many calamitous possibilities out there, how could I *not* have one? Or at the very least, some minor and heretofore undocumented variation?

Don't go there.

I took it as a sign of recovery that I didn't go there. My mood was unquestionably on the mend, even if my nose wasn't. Was it the SSRI I'd been taking since Christmas? Was Lexapro finally kicking in?

20

NO QUICK FIX

O NE REASON WHY the nose has been neglected is that smell dysfunction is relatively rare. In humans, the sense of smell actually *outlasts* vision and hearing. Before you counter that your grandmother sees just fine but can't smell a thing, think about those glasses she's wearing. While the nose makes a dandy perch, nothing gives *it* a boost later in life. We're so tuned in to vision that we don't see bifocals as signaling the onset of our inevitable (if we live long enough) blindness. Note the use of the word *see* in that sentence. We esteem vision so highly that we've come to say *see* when we mean "think."

Hearing begins to deteriorate earlier than olfaction too. A 2008 study found that 8.5 percent of people in their twenties have some hearing loss.

You have to search hard to find an anosmic that young. Two percent of the population below sixty-five is essentially smell-blind; a quarter of that 2 percent was born without smell. Smell declines precipitously in old age; half of all eighty-year-olds report diminished smell. An estimated half a million people annually see a doctor about olfactory problems.

Gerontologists invariably blame depression for their patients' poor appetites, even though depression is actually less common in the elderly than in the general population—unless the elderly person is anosmic.

It's not simply because hearing and vision problems show up earlier in life than smell problems that medical science has made a sizable investment in hearing aids and eyeglasses. Bias against smell is expressed by the AMA in its impairment assessment. Deafness is taken so seriously that parents who've passed deafness on to a child feel pressured to give the child the gift of hearing—a cochlear implant. (Understandably, many such parents want their children to hear the way they do, through sign language.) A ninety-five-year-old California man who'd gone deaf managed to persuade his insurance company to cover his expensive implant because he said it would improve his health (lower his risk of a stroke or heart attack) by assuaging his loneliness.

The loneliness felt by anosmics is not viewed as detrimental to their health. It's not viewed at all. Pundits aren't squabbling over whether a nose implant should be covered by insurance. Moreover, anosmics keep their problem to themselves. This is easy when no one else is inconvenienced by it. Easier still when even the anosmics don't know why food tastes awful and why they keep forgetting to turn off the gas. They suffer in silence and often in ignorance. Electronic noses do exist, but these contraptions don't look anything like a real nose. They're used to detect rotten meats and toxins in commercial settings. Few if any scientists foresee a time when a person with smell dysfunction will be able to strap on a bionic nose and smell again.

And that's not just because smell is a second-class citizen. There's a significant technical obstacle. The pathway for smelling, as we've seen, is widely distributed, with intricate wiring deep inside the brain, not just at the periphery. This makes it a far more complicated sense than hearing. Moreover, those cochlear implants used for deafness don't actually restore sounds in the familiar way. Patients describe the new sounds as highly unpleasant at first; their meanings must be learned with the help of a computer program, much as a blind person has to learn Braille. With meaning comes appreciation for this new auditory world. The ultimate challenge is to develop at least a little enjoyment of music.

The same is true for people who lose their eyesight at a young age. Recently, scientists unveiled the world's first electronic eye. When asked if it could cure blindness, they readily admitted their "eye" wasn't for humans but for digital cameras. It's one thing to treat glaucoma (once the leading cause of blindness) with drugs; it's quite another to replace an organ that has developed organically over the years in lockstep with its owner's experience. A man named Mike May, blinded by an explosion at age three, received a brand-new cornea in middle age. Years after the surgery he still couldn't recognize his kids. Mother Nature always puts a negative spin on novelty, and, sadly, May "saw" the world as ugly and chaotic, just as I smelled the world as ugly and chaotic when my nose was disjoined from my mind.

Joseph LeDoux explained that "in order to see an apple, instead of a roundish, reddish blob, the various features of the stimulus, each processed by different visual subsystems, have to be integrated." The cognitive aspects of pattern recognition were no longer available to May. His brain had long since rewired itself to accommodate other sensory inputs. May had been a champion downhill skier when he was blind; his new eyes destroyed his formerly superhuman balance and uncanny navigational skills. Now he couldn't manage the bunny hill.

For this same reason, science is more likely to come up with a Hummer that runs on air than a truly bionic nose. An electronic nose that detects the odor of rotten meat is a different technology altogether, as different from the human nose as a computer is from the human brain.

Smell dysfunction did spend a decade or so on the NIH agenda, partly because chemotherapy was destroying taste and smell in cancer patients and partly because new research had found that inhaled chemicals could cause disease. The discovery of the smell genes shifted funding priorities from smell treatment centers, like Richard Doty's, to genetics labs, like Richard Axel's. Molecular biology seemed more likely to solve "important" problems. The direction of smell research was quietly changed; some NIH-funded smell and taste centers were forced to close, and most other labs now rely on private investors.

Modern medicine is about treatments and cures. ENT doctors have ears and throats to worry about, as well as sinus conditions that may or may not cause smell problems. I was very lucky to have landed in Dr. Cushing's office. Few patients come in complaining to an ENT that their world smells like rotting flesh. Some internists and family practitioners aren't even aware that *anosmia* is the word for smell-blindness. Or that smell-blindness is something people get, like a cold, and that they can get it sometimes *from* a cold, sometimes from a cold remedy. Or from falling off a bench. Or from a car crash. Or from cancer. Or from Alzheimer's disease. More often than not, people with smell loss or distortion are referred to either a neurologist or a psychiatrist.

Many scientists have the same lack of awareness about smell. Richard Doty admits that he "sort of fell into" his line of work. He adds that he wouldn't make the same mistake again.

Richard Axel says he doesn't think of himself as a smell scientist but as a gene splicer. Linda Buck recognized early on the medical-breakthrough implications of her discoveries, though curing smell dysfunction itself wasn't (and still isn't) on her to-do list.

In a curious twist of fate, the 1991 discovery of smell genes in humans was a setback for people like me, as olfactory science shifted its focus from studying human smell dysfunction to investigating the brains of anesthetized rodents, fruit flies, salamanders, and frogs. At the same time, high-tech tools like the gas chromatograph, the mass spectrometer, and the two-photon microscope accelerated the shift in emphasis from species behavior to cell biology, and functional MRIs made top-down research just as valid a method of inquiry into olfaction as reductionism. The high-tech tools also spelled trouble for smell psychologists: no need for B. F. Skinner's elaborate mazes when you can watch cause and effect in real time inside the brain itself.

Richard Doty opened the first smell and taste treatment center funded by the National Institutes of Health. In 1999, he was named one of the two thousand most outstanding scientists of the twentieth century. He was awarded the prestigious Sense of Smell Award in 2000, a distinction he shares with Trygg Engen, Gordon Shepherd, Richard Axel, and Linda Buck. The day after my visit to the Axel lab in New York, I took a train to Philadelphia. Doty's smell and taste treatment clinic consists of three small rooms in a classroom building, each one furnished with plastic chairs for patients, and bookshelves for Doty's writings and research articles.

I told him my story. He was well aware of the Zicam controversy and agreed that the FDA should do something about such products. He suggested some people I should talk to about smell, including a scientist at Cornell with Parkinson's who was investigating possible links between the disease and inhaled toxins. Doty sent me home with a free UPSIT (the University of Pennsylvania Smell Inventory Test, which rates the severity and specificity of smell loss) kit. It's still unopened—did I really need to confirm what I already knew?

Several months later I e-mailed Doty with a question concerning a treatment for congenital anosmia that I'd heard about. I mentioned a certain doctor who ran a smell and taste center on the East Coast, ostensibly no different from Doty's own although private and not underwritten by the NIH. Doty replied with a curt e-mail. Yes, he knew this man. The subtext of the message was clear: *stay away!*

This East Coast doctor, Dr. R., as I'll call him, claimed to have come up with cures for both congenital anosmia and phantosmia, tackling the latter through a procedure that effectively puts the olfactory bulb out of service, stopping not only bad smells but all smells and ensuring that the patient won't ever smell anything again. To "cure" congenital anosmia, Dr. R. surgically transplants receptor cells from the tongue to the nose's receptor sheet and then exposes them over and over to certain strong odors until the relocated cells get the hint and —all hail plasticity—turn into odor receptors. Doty fumed: "Only a highly skilled neurosurgeon has any business fiddling around with brain tissue."

What about the testimonials from over-the-moon beneficiaries of the congenital-anosmia breakthrough? I asked in another e-mail. Doty patiently picked them apart, noting how often the restored "smells" could be attributed to sensations from the trigeminal, not olfactory, nerve. "I can smell mustard!" one said. But how could a person born with no sense of smell know the difference between an odor and the sting of a chili pepper? He pointed out that wishful thinking is a powerful force and that most of the patients spoke of their hopes for the future "if they just kept up the 'training.'"

I was reminded of Oliver Sacks's story about the man who willed his sense of smell back (or thought he had), and of Elizabeth Zierah, the anosmic who wrote that she was willing to try anything to restore her sense of smell. She finally scheduled a risky reconstruction of her sinus cavity. The procedure was dangerous and had a low success rate. She didn't care.

Dr. R.'s abstruse accounts of his experiments had been picked up by second- and third-tier journals (and then by the news media) that inadequately vet their articles, Doty complained. Such quacks also distribute their articles and testimonials on the Internet and any other place that is open to all comers. They refuse to return phone calls from serious reporters, he said, and suggested I give that a try.

Sure enough, my communication with Dr. R. ended almost before it began. After a few mysteriously missed phone connections, he told me to stop bothering him until I'd read his articles thoroughly and knew something about smell. I was wasting his precious time.

Where had I met this man before? Oh yeah, in *The Wizard of Oz*. Remember how angry the "wizard" became when Dorothy asked him unanswerable questions, and how he lost his temper completely when her little dog, Toto, pulled down the curtain to reveal that the mighty and terrible wizard was a fake?

One day shortly after this incident, I saw an online ad for

the latest "cure" for smell dysfunction—neurofeedback. The pitch went like this:

> If you suffer from loss of smell, ask your doctor about the use of neurofeedback in your treatment and rehabilitation. Once [you're] approved for neurofeedback treatment . . . the therapist will ask that you sit quietly with your eyes closed. Placing electrodes onto the head, your therapist will send very small doses of electromagnetic signals to the brain, promoting flexibility and stimulating the brain to perform normal functions again.

These people were casting a wide net. Neurofeedback's alleged "success in managing complications of epilepsy, ADHD, autism and even juvenile offender cognitive processing" virtually guaranteed success in tangentially related areas. I was (briefly) tempted to give it a try for my insomnia and migraine headaches. As to smelling, neurofeedback promised that anyone suffering from "brain complications in the loss of smell may soon find their nasal receptors are re-activated." Or . . . maybe not.

Alan Hirsch, the man who invented the diet aid called Sprinkle Thin (repackaged as Sensa in 2006), is the founder of the Smell and Taste Treatment and Research Foundation in Chicago. A fast-talking obsessive (it takes one to know one), he wears V-neck cashmere sweaters and drives a white Jaguar. Boundaries? Don't try reining him in; it won't work. He might dream up a cool experiment this morning and have it fully funded, staffed, and ready for liftoff this afternoon. He has written several books on smell for the popular audience, including *Scentsational Sex* and *What Flavor Is Your Personality?* He has no qualms about turning his scientific discoveries into fodder for *Oprah*.

His colleagues consider him a respectable peer, if a trifle over-zealous.

Richard Doty, who is as respectable as they come in this business, calls Hirsch (while grinning) "a very creative guy." Doty's own operation competes for dwindling grants from places Hirsch probably doesn't bother with, but because both spend most of their time on health issues related to olfaction, they were quoted at length in the same 2007 *New York Times* article about products being developed for the early detection of Alzheimer's through smell.

Nick Kokonus, co-owner of the Chicago restaurant Alinea, told me his wife had lost "most" of her sense of smell in a car accident and was working with Hirsch to get it back, with some success. This obviously piqued my curiosity. What sort of person was this Hirsch? For one thing, he is board certified in both neurology and psychiatry. He got into smell because he had some ideas he wanted to explore. His original area of interest, human behavior, perhaps helped persuade him that he could parlay his findings on smell into a small fortune in popular books and health products.

I arranged to meet Hirsch at a coffee shop in the northern Chicago suburb where he lives. It was almost April now, six months and counting since I'd lost my sense of smell. I waited fifteen minutes and was about to phone him to see if I'd gone to the wrong Starbucks when a small, wiry man swept through the door carrying a bulging leather briefcase.

"Would you like me to tell you about my new diet aid?" he asked not thirty seconds into our interview.

I wondered if he'd noticed my pants were a bit snug. Hirsch has conducted hundreds of studies to get at the causes of both obesity and food preferences (the latter research funded by companies trying to spot trends in consumer behavior). Just as he doesn't mind selling his expertise to manufacturers that make things like corn chips and room deodorizers, he isn't picky about where he's published. This accounts for the foot-thick stack of papers he somehow managed to extract from the briefcase and flop on the desk in the library cubicle we wound up in after Hirsch decided that the crowded coffee shop was too noisy for constructive conversation. I wondered if he wore cologne. He was so perfectly dressed for cologne, down to the gold chain on his wrist and the slicked-back hair, that I could almost catch the scent. English Leather? Brut? Are those even sold anymore? I decided not to ask about his cologne (which of course I couldn't *really* smell), even though this demonstration of the power of imagination to conjure up smells would have captured his interest. He might get other ideas. Ideas were clearly something this man got as easily as most people got colds.

An interest in why food tastes vary from person to person had launched another Hirsch project, one sure to grab a headline or two, maybe even land him a book contract or at the very least an article in *Chemical Senses* (a journal put out by Oxford University; its readers are about equally divided between academia and industry).

Hirsch had decided to test top chefs' noses. Surely they'd all be superb smellers. Hirsch gave them each a smell IQ test. He wanted to know how well the chefs smelled compared with ordinary people. Then he'd find out how (and if) they'd trained their exceptional noses. Did any of them claim to have been "born" to smell?

While not all human noses are created the same on the outside — thus the prevalence of nose jobs in a culture obsessed with image — we each have essentially the same innate sense of smell. Exposure to certain odors over and over is what sets professional noses apart from ordinary ones. Wine tasters, for example, train their noses with the wine aroma wheel, developed by chemist Ann Nobel in 1984. The wheel supports the component smells of a certain vintage with words and pictures as memory aids.

Hirsch wasn't trying to overturn the training theory, just flesh it out. He knew it was unlikely that any of the chefs would display the olfactory equivalent of perfect pitch. But he was surprised when four out of ten chefs were unable to tell simple everyday smells apart. Only two were excellent smellers, and two others couldn't put a name to much of anything. This confounding result reminded Hirsch that you don't have to identify an odor by name to appreciate it and combine it effectively with others. Beyond that, he concluded that maybe just as important as a well-trained nose is mastery of time-honored cooking techniques. Other essentials: buy the freshest possible ingredients, understand food chemistry (especially the effect of temperature), and hire a fabulous decorator.

As for the chefs who flunked the test, Hirsch could only speculate that fame and fortune had gone to their heads—or rather, up their noses. "This is a glamorous profession if you make it to the top. These guys"—there was not a single female in the group—"are young and suddenly pulling down six figures. There are temptations too strong to resist."

The last time I saw Hirsch, in the summer of 2008, was on TV. Caroline called me from Madison to tell me "that nose guy in Chicago" was being interviewed on ABC's *20/20*. It seems the show's producers had done some research of their own. When the diet aid was run by an authority on medical scams, he declared Sensa "just another pet rock."

21

GOING AFTER ZICAM

There's nothing new about homeopathic medicines. Until the twentieth century, doctors offered their patients pretty much anything they thought might bring relief, even if the only active ingredient was grain alcohol. Turpentine is a key ingredient in what is still one of the most trusted names in homeopathy: Vicks VapoRub.

My mother was born one year before the Spanish flu pandemic. Her mother had been a worrier, terrified of infectious diseases: Diphtheria. Whooping cough. Scarlet fever. When she was a girl, Mom had been put to bed if she had even the mildest case of the sniffles. Some of her loveliest memories were of those frequent quarantines when my grandmother plied her with presents and treats and rubbed Vicks on her chest. Its blend of camphor, menthol, turpentine oil, and eucalyptus spiced with nutmeg, cedar, and thyme is as soothing today as it was in 1880, the year a North Carolina pharmacist came up with the pungent formula to treat nasal congestion and named it after his brother-in-law, a prominent physician. Vicks VapoRub is made in Mexico now. The magic formulation hasn't been tampered with in all these years, although it's currently marketed as a cough suppressant rather than a decongestant.

There's no law against selling a homeopathic remedy to make people feel better, even if its effects are purely psychological. When plaintiffs in the Zicam lawsuit asked for an investigation, they received little satisfaction from the FDA. The agency didn't see how a nasal spray consisting of organic compounds could be any different from a skin potion sold as a cough suppressant. They seemed to work the same way, by going up the nose. What could be the harm in that?

Winter was winding down. Minnesota noses are trained to catch the faintest hint of changing seasons: Minnesotans begin smelling spring in the dirty snow melting on the sidewalk as the drifts recede; in rotting newspapers still rolled up and fastened with rubber bands; in cardboard soda-pop cartons left behind on recycling pickup day; in a leather glove, a wool mitten, an old tennis shoe, last year's oak leaves (leaf mulch now). Yes, spring is just around the corner.

In March I was asked to give a talk on late-blooming perennials to a garden club numbering about a hundred women; I decided to do it. Another milestone. Back in the fall I'd turned down several opportunities to expound on the wonders of spring bulbs because I couldn't face an audience. This time I accepted.

A fragrant anemone popped onto my slide screen. Tears welled up in my eyes and the story of my anosmia came burbling out of my mouth. The women were polite, if bewildered. I could see them whispering to one another. Eventually I got back on track. Where were we? Oh, yes, late-fall bloomers.

A few hands shot up at the end when I asked for questions. Every other person, it seemed, knew someone who knew someone . . . One woman had lost her sense of smell at about the same time I had. Until I mentioned Zicam, she had suspected but never openly accused the nasal spray in her medicine cabinet of committing the crime. She felt so stupid, she said, falling for the hype that the spray would help her cold. When her nose stopped working, in an odd way she thought it served her right for using medicine in the first place. Since when was a simple head cold so terrible a person couldn't just grin and bear it? Typical gardener, I remember thinking, someone of a certain age and era, old enough to have planted a victory garden with her parents during World War II.

We exchanged phone numbers. When I got home, others called, people who wanted to know what I knew about Zicam, anosmia, phantosmia, and the connection between smell and mood. Some said their doctors had dismissed their concerns about poor smell and taste with the dubious explanation that smell loss was an inevitable consequence of aging, like aching joints and decreased hearing, although of course untreatable. The untreatable part is true, but it's also true that while half of all eighty-something seniors have significant smell loss, smell function doesn't peak until age thirty-five, and decline is barely measurable for many, many years after that. Some people's noses never slow down at all.

When smell does diminish, medications are often the cause. I'm not talking about Zicam. An Australian study assuaged any lingering embarrassment I'd had for suspecting that a blood pressure drug could be messing with my nose. The people in the study who'd been taking drugs to lower blood pressure and cholesterol suffered significantly more olfactory damage than those who were pill-free. Even SSRIs, the family of antidepressants that I was taking, have been suspected of reducing smell acuity.

Such research relies on personal testimony. Smell scientists still don't know enough about how odor signals are processed in the brain to be able to speculate on why certain seemingly random medications diminish smell. The default position is, of course, to toss the whole issue (along with disorders like burning mouth syndrome) into the dumpster labeled *psychosomatic illness*.

I told some of the women about a website called nosmell .com, where victims of smell dysfunction compare notes. I continued to visit it whenever I needed company. One recent post brought back all the misery of the holidays. I might have written it myself.

> I just recently started having this rotten, rancid smell and/ or taste. It started after a head cold in which I used Zicam. It is driving me CRAZY! The smell/taste never stops. I also worry about my own body odor. I can't even tell if I have any! This is a nightmare.

The post put me over the edge. Before, I had been too distraught to focus on why this had happened to me. Losing confidence in one's sanity is not conducive to waging a legal battle. Now I searched for more recent information. Finding it wasn't hard. An online blogger described losing his sense of smell after taking Zicam for a cold just before Christmas. He was awakened in the middle of the night by the smells of burning coffee and cigarettes. He went to several doctors, had MRIs and CT scans, "was poked, prodded, had blood taken . . . all without any evidence of any underlying tumors or other possible causes of the problems." Like me, he finally went to an ENT, who knew immediately what was wrong and told him to get an attorney. Beyond that, the doctor could do nothing to help him. A colleague confided to him that he too had lost his sense of smell after using Zicam and the loss was so devastating he went into hiding and told no one. The blogger was determined "to take the absolute opposite tactic. I'm telling everyone . . . and now he is too."

He concluded by warning that all Zicam products, including the new control-tip spray, were worthless or worse. There are now more than a dozen Zicam variations, each one targeting a different nasal annoyance. This strategy has expanded Zicam's reach by increasing its shelf space. Now a sea of orange greeted me at Walgreens when I picked up my FDA-approved prescription drugs. The rows and rows of bold orange boxes shamed me for failing to protest my injury in court or even call the FDA to complain.

Many of these products piggyback on other, more efficacious remedies. Zicam tablets contain vitamin C. Zicam gels, swabs, and sprays contain aloe, which soothes irritated membranes. (Are those membranes irritated by the cold or by the remedy?) There's a Zicam cure for sinus pain, coughing, allergies. Zicam products don't just prevent colds, they relieve them when they're full-blown by means of a decongestant that is probably no different from any other decongestant. During the time I spent researching and writing this book, sales of Zicam products soared from $6 million to $100 million. Sales were projected to grow from the 8 percent annual rate posted in 2008 to 10 percent per year by 2009, despite, as Matrixx Initiatives reminded prospective shareholders, an otherwise horrendous economy. Twelve million dollars—the amount the company spent to settle the lawsuits—is a small price to pay to stay in such a lucrative business.

In 2006 the *Washington Post* published an article called "The Men Behind Zicam." It described the "unusual backgrounds" of Zicam Cold Remedy's two inventors, Robert Steven Davidson and Charles B. Hensley. Both hold patents on the chemical formulation bottled in Zicam Cold Remedy. Davidson said he first met Hensley at Cleveland Chiropractic College in Los Angeles. Davidson received a bachelor's degree from a "virtual" university in New York and earned his PhD from an unaccredited university in Spain that has since been closed by Spanish authorities for legal violations. Davidson declined to discuss Zicam's safety problems with the *Post* reporter. He also claimed to be unaware that one of his alma maters had gone under. He said he'd sold his interest in Matrixx Initiatives and had launched another company.

Hensley's recent track record is even more troubling than Davidson's. While his doctorate (in physiology) is from the University of Southern California, he now runs a pharmaceutical company that received a warning from the FDA in November 2006 to stop marketing its antiviral drug Vira 38, which the company claimed was effective against bird flu. The FDA threatened legal action and "seizure of illegal products" if its letter went unheeded. "Hensley did not respond to e-mails or telephone calls," according to the *Post*.

In 2004 Matrixx Initiatives settled the class-action lawsuit against it for $12.5 million; each plaintiff received about twelve hundred dollars (the settlement agreement required that they not reveal the precise sum). Some plaintiffs said they felt let down and angry about the case's resolution and said they wouldn't have settled had they been given a say in the matter. Press coverage was largely confined to the *Washington Post.*

If Matrixx Initiatives had messed with any other sense — had its product, say, caused an unceasing din in the inner ear — the outcry would have been deafening and the company out of business in no time flat. A mere whiff of a serious side effect can sideline even the most useful medication, and this product was a nasal spray of questionable benefit. In addition, it contained a chemical that had, decades earlier, damaged the olfactory cells of hundreds of children. In Toronto in 1937, a study was conducted to determine if an intranasal zinc solution might prevent polio. Many of the five thousand children given the spray got polio *and* lost their sense of smell. For a while, doctors thought anosmia was another symptom of the dread disease.

Matrixx spends far more on advertising than on legal costs. Particularly galling to me were the ads claiming that "doctors recommend Zicam to their patients" and the dozens of testimonials from happy Zicam customers. Rush Limbaugh, who promotes the product and whose show is sponsored by its manufacturers, apparently never gets colds. (The reformed user of illegally obtained nonhomeopathic narcotics does wear a cochlear implant, though; it saved his career.)

That polio connection was problematic. Matrixx promulgated its version of the story. The polio-prevention treatment had contained a different zinc compound, the company said. Zinc gluconate (a.k.a. zincum gluconicum), the compound in Zicam's product, was harmless. It's zinc sulfate that causes trouble, Zicam's medical experts said. Medical experts paid by the plaintiffs' side didn't agree.

Moreover, there is no evidence that any compound of zinc works better against colds than it does against polio. Various theories do explain why Zicam Cold Remedy might *seem* to work, although not in the low doses Matrixx Initiatives suggests. And of course, a massive dose can do plenty of damage. Zinc gluconate has been shown to destroy the olfactory sheet in mice. One mysteriously ill woman with extreme fatigue, high blood pressure, low thyroid levels, muscle soreness, mental confusion, and balance problems had been using large amounts of a denture adhesive to keep her dentures in place so that she could indulge in the only pleasure left to her: eating. It turned out that she'd overdosed on the zinc in the adhesive, and that was what was *causing* her problems. Her story, "Fear of Falling," was told in Lisa Sanders's September 6, 2009, health column in the *New York Times Magazine*.

Why do so many people swear by Zicam? They want to believe it works. Maybe users credit the spray instead of the body's immune system when a cold is snuffed out or is milder than normal. What makes this all add up to the perfect product is that Zicam is innocent until proven guilty. Unlike allopathic drugs, such as, say, Lipitor, homeopathic drugs don't have to be proven therapeutic. They don't even have to be proven safe.

The antiregulatory attitude in Washington, D.C., has been as big a help to Zicam as Limbaugh's endorsements, and a lot cheaper. When a Dow Jones Newswire article in February 2004 reported that a spate of consumer complaints about Zicam had prompted the FDA to look into the product, Matrixx Initiatives insisted that Zicam adhered to FDA guidelines and restrictions. Eventually the FDA coughed up an ambiguous statement implying both that there had been an inquiry and that there had not. It considered zinc gluconate "generally" to be safe "although this does not constitute a finding by the FDA that the substance is a useful dietary supplement."

It turns out that several hundred people had complained (to

no avail) to the FDA that Zicam had destroyed their sense of smell. The number of plaintiffs in the class-action case alone should have tipped off the agency that this was not a frivolous lawsuit. For every victim who signed on, there were probably thousands who never made the connection between Zicam and their anosmia.

Forget about anosmia then, I wanted to scream when I found the FDA statement online. Think about all the other neurological illnesses it could cause. Zicam nasal spray makes a beeline for the brain by way of olfactory axons that have connections to the limbic system and the higher brain. Consider medical science's growing concern about all the chemicals that go up noses and that can possibly trigger chain reactions with catastrophic long-term consequences. Researchers barely understand how synaptic responses occur, much less what might set them off. Does anyone really want to endanger the brain for the sake of small-government ideals and big corporate profits?

Consumer Reports magazine reviewed the Zicam literature in 2007 and concluded that in view of the murky research and ongoing complaints, consumers with colds coming on were far better off drinking lots of water and going to bed than shooting a potentially toxic substance up their noses.

I spoke with a lawyer in Arizona who worked for the firm that had settled the original case. The lawyer said cases were pending in other states. A new swab applicator in the product was supposed to eliminate any possibility of overspraying. (It too is now being linked to anosmia.) Matrixx Initiatives was still insisting its product had been safe all the time—not just safe but effective—and that zinc gluconate dissolved into zinc ions and gluconate, both of which were "naturally occurring compounds" found in all human tissues. The company still had no comment on the original plaintiffs' almost unanimous assertion that the product had caused "a strong and very painful burning sensation," other than to imply that the directions on the package were meant to be read.

Those directions advised pumping the gel into each nostril but not sniffing. Brilliant. Just the sort of thing that makes the silent majority say to themselves, *How silly of her,* while a tiny minority of whiners like me wonder, *How on earth was I supposed to do that?*

Caroline had never taken her nose for granted. She actually used her sense of smell to *create* good memories. She was conscious of the smells of things she loved and their soothing effect on her emotions — the special fragrance of our dog Mel's skin, the cat's breath, even the smell of our kitchen cabinets with their aroma of old shellac and the molasses that had been sitting inside one of them so long the glass bottle had years ago begun leaching its sweetly pungent scent. She was very particular about soaps and kept a scented candle lit when she was studying. It calmed her down, she said.

In her heart she held fast to the belief that the confusion in my nose would clear up and all would be well again. I, being the only one in the family — the only person on earth, for that matter — who knew the truth by virtue of hard evidence, was for once the realist and not the dreamer. My job was to listen patiently to my family's well-intentioned opinions, remind myself that they loved me and could not bear to give up all hope of my eventual recovery, and go about the business of adjusting to life without smell on my own. So it is, I realized, with anyone who's been diagnosed with a life-changing affliction. In an odd way, my not challenging the opinions (hopes) of others made it easier for me to accept the truth. Love is unselfish. Personal adversity proves that.

My all-consuming hope was not that my nose would wake up but that the anosmia would not change me or my life in any way. I would get used to this. It was no big deal. My daughter protested this way of thinking, even though it was the mantra she'd given me—"Put one foot in front of the other"—that made the goal seem within reach. My low expectations irked her; it seemed wrong somehow. False. Attitude is *not* everything, except maybe in self-help books, she said. Why set yourself up for failure by pretending that a person can and *should* control emotional events? Why deny their power to hurt? "Losing your sense of smell is not a small thing."

Smell used to ground me in the here and now. It took the edge off my essential solitude. It challenged my irrational (or not) fear that reality is unreliable and can slip away at any moment. Certain smells are ravishing and others foul, but all of them possess an animal component that is absent from sight and hearing. You can't overthink a smell. It's there whether you want it or not, having its way with you, like music, but more potent for its subtlety, its immunity to reason, how it affects you without your knowing it, how it makes things real on their own terms. Makes *you* real in a way that has nothing to do with you.

Based on my research on olfaction, I was beginning to believe that treatments for anosmia would come along someday. Would this SSRI, Lexapro, help me think positively again? It already had. I hadn't taken a tranquilizer in weeks.

Besides, thinking *clearly* was enough for the moment. Positive thoughts could wait. I felt blessed that I could read a recipe, enjoy a good book. I could converse without having to shout over the static. Understanding the brain, even just a little, was bolstering my confidence. I would get over this olfactory loss. It was comforting, even humbling, to figure out that the emotional brain repairs itself just as the body does. If I could just stop picking at my wound, it would heal.

In late March a carpenter showed up at my house to install a new countertop. Just because I couldn't cook anymore didn't mean I didn't relish the prospect of a kitchen remodel. The carpenter, his friend—an ex-Marine named Larry—and I measured and cut heavy granite slabs with a diamond-blade saw and then wrestled them into place, all of us wearing 3M facemasks, at my insistence.

This prompted Larry to make a startling confession. He'd been anosmic for thirty years. Then his sense of smell woke up. He'd always figured his nose had gone on the blink when he'd lost most of his hearing. (He was a Marine pilot in Vietnam and a bomb had gone off in his plane.) As it turned out, he had nasal polyps, and after he had surgery to remove them, his smell came back, he said.

He was amazed that no one had ever told him they might be affecting his sense of smell. "Not that I ever asked. I probably had twenty percent [of smell], like I could figure out what something was if I held it close. But food? Nah. I couldn't have cared less about eating.

"When I got my nose working again I put on twenty pounds right away," he said, happily thumping on his slight paunch. "It's great!"

As we worked on the counters, Larry talked on and on about how wonderful it was to smell. "I won't ever get over it," he said. He couldn't wait for spring to come every year. He used to look forward to the fishing opener. Now he spent all day walking

around smelling lilacs. He'd become something of a smell con-
noisseur, he said. He no longer slipped out the front door when
his wife looked like she was about to ask him for a hand in the
garden. He couldn't wait to help. He'd decided that lilies were
the most fragrant garden plants and that his wife's white Casablanca lilies "put all the rest to shame."

Another thing he'd noticed: He'd always struggled with depression, even tried to kill himself once after he left the military. He'd just assumed it was related to serving in Vietnam.
He'd hit bottom when the Marines discharged him because of
his hearing loss. "But if you want to know the truth, getting rid
of those polyps was just about the best thing that ever happened to me," he said.

I was happy for Larry. Really I was. I was doing just fine
without smelling. I was riding my bike again. I was even gathering ideas for a spring issue of the *Garden Letter,* and when a
food or garden magazine came in the mail, I did *not* toss it in
the recycling bin.

22

SPRING

IN SPITE OF ANOSMIA (or because of Lexapro?) my range
of concern was widening. My worries had begun to resem-
ble other people's again: Climate change. The casualties in Iraq.
At the same time, previously all-consuming issues such as Caroline's homesickness (would she drop out of college?), Cam's
acid reflux (was he headed for cancer?), and the likelihood of
my going insane were being restored to more appropriate positions in the constellation of concerns that needed my attention.

This made room for addressing pleasurable topics like
painting the kitchen (I'd wear a respirator), *finally* reupholster-
ing that faded and filthy wing chair, and even planning a bike
trip in Italy. Travel didn't seem the least bit terrifying. Nothing
seemed terrifying, with a single exception.

Spring. How I dreaded its coming. I was like a five-year-old
fretting about a flu shot. The pain of that comes and goes in a
split second, any lingering soreness swamped by the euphoria
of knowing that the worst (anticipating the dreaded prick) is
over. "That wasn't so bad, was it?" the nurse always says. The

child blushes and shakes her head no. It's not the pain that's hard to take but the worrying.

My situation was a little different. A flu shot—even the flu—comes and goes. What was worrying me was how I'd react to missing something that I hadn't given much thought to before. Much as I love the smell of my garden, it was its visual aspects that occupied my thinking brain. Color combinations, unsightly weeds, problems of proportion and scale. I'd planted my herb garden close to the kitchen mainly for convenience. Yes, I used to think of myself as a smell aficionada, but having fragrance close by had been an afterthought.

When I went outside in the spring to rake the leaves off my perennial beds, would the once reliable adrenaline rush kick in as usual? Would the hosta's sharp green tips strike me as endearing, the way they always had, without the aroma of the soil, moist and pungent, as the ice melts and allows it to breathe again? Would my garden still captivate me without the olfactory associations of past springs and summers, farms and forests, storms and drought—without any connection to memory?

Or would every scentless rose petal and leaf remind me that I couldn't smell it? When the grass was mown for the first time and the lilacs were in bloom and the herbs just beginning to leaf out and I couldn't smell any of it, what would that feel like? Would planting, weeding, and pruning be no different from scrubbing the floor? I was sure the healthy parts of my limbic system, the amygdala and hippocampus, were in mourning and passing their sadness on to me.

For Cam it all boiled down to this: How did I get through the day without looking forward to dinner? Lots of my problems, imagined and real, struck my husband as silly. Not this one. "What does food smell like?" he kept asking. I told him I wasn't sure, and that was part of the difficulty. He persisted. "I don't mean emotionally, but, you know, what's it like?" My wonderful husband really did want to feel my pain. He would gladly have given up one of his olfactory bulbs so that I could smell again, maybe even both of them. But you can't donate an olfactory bulb like you can a kidney.

I started inventing sensations just to satisfy his curiosity. "Anosmia is . . . Minnesota in January when it's twenty below and the skies are gray and everything is dead." (The truth is, Minnesota in January has a distinctive tang—sharp and fresh.) One day I said, "Anosmia is like air conditioning. Can you imagine that?" He knew why I'd chosen that analogy: I was sick of being asked this impossible question. We argue a lot about air conditioning. I always want to open more windows, and he (the man who loves tropical weather) wants to "turn on the airco." "Anosmia is like that," I told him. "A void, a vacuum, a three-dimensional world abruptly reduced to two."

"Look, at least you're not crazy."

I replied out of habit that without smell, it was just a matter of time before I became confused and delusional.

"You've always been confused and delusional," he said.

We both laughed, Cam at his own hilariousness, I at the person I used to be.

"But you *can* smell, I know it!" Cam shouted from behind the refrigerator door. "Where the hell is the mustard?" He fished around for the Grey Poupon, found it, whipped off the cap, and shoved it under my nose. "Here, smell this." Exasperated, I took a sniff. He was right. Sort of. But was I really smelling it, or merely remembering what mustard smelled like? Anyone who can imagine she's experiencing the symptoms of MS is surely capable of conjuring up the sharp aroma of mustard.

Cam was jubilant. He began madly foraging in the spice rack for cloves, cumin, garlic. My husband and I both love to cook, but as I got more involved in gardening, my role became that of supplier of fresh herbs and vegetables for his gleaming sauté pan. He was disappointed that he might lose his culinary playmate. The long meals we'd always loved to linger over, our conversation drifting easily between politics and the movies to whether the wine goes well with the scallops and if the scallops might have been left a bit too long in the pan — all this shared pleasure was gone, zapped by a nasal spray. His having brought the evil potion into our home was another reason he was fighting this diagnosis. He insisted it was the cold virus that had done it, "if in fact you actually have lost your smell." How could he be sure I wasn't just, well, making this up?

On my next visit to Dr. Cushing, Cam came along. The doctor was pleased to hear that I seemed to be able to detect some scents, however faintly. (Cam was sure of it.) This was definitely a good sign.

Orange slices. Onions. An overripe banana. Cloves. One by one they were passed under my nose for Cam's smell test. The cloves convinced me that maybe I did have some smell left. This boosted my spirits, until I learned that it was the trigeminal nerve, not the olfactory nerve, that was "smelling" them.

Every day that passed without improvement reduced the chance I'd smell again. Things settle in. Change gets harder. Surely brain cells are no different from habits in that regard. If the stem cells *were* regenerating, I wouldn't feel anything up there telling me so. No slight tingling sensations. My nose wouldn't itch. One day I'd smell something and the odor would strengthen. Or it wouldn't. After an injury like mine, if something good is going to happen, it should happen within about six months. People who lose their sense of smell after a brain injury are given a year of hoping before they are urged to accept their disability and move on.

The calendar said the month was April. I looked to the trees to confirm this, and it appeared to be true. Most leaf buds had shed the hard brown shells that protected them from the cold. The trees' dark branches wore green specks. My perennial beds were piled so high with straw and leaves it was hard to know what might be rustling underneath.

I should have removed the mulch by now; something had told me to stay away. What I discovered could only lead me back to despair. The tall oaks that stood at each corner of our yard like giant pushpins, seeming to anchor it in place, had yet to surrender to spring's siren call. They would help me, these magnificent stoics. If they could weather a century of Minnesota blizzards, surely I could get through a scentless spring.

Put one foot in front of the other.

I read the newspaper, sipping tasteless coffee. Most mornings the buoying elixir caught me unawares, just the way the aroma hitting my nose always used to, and while I didn't have smell back, I had the rest of me. Sometimes I even imagined I'd caught a hint of French roast. It amused me that I seemed to be repeating the story of the man Oliver Sacks wrote about, the one who'd convinced himself that he could smell pipe tobacco. The gentle persuader in his case? Not caffeine, but nicotine.

At nine or ten each morning I made the climb from the kitchen to the attic, with Mel by my side, begging me to toss him the ball for a few minutes before I got down to work. Every day the climb up to the attic got easier. The ball-tossing became less perfunctory and more fun. With Mel curled up on the floor beside me, my office felt cozy and familiar. I was not alone. I didn't feel detached. Sometimes I lifted him up and we'd cuddle in the chair. Alex had taught him to kiss; I used to find it repugnant, but now, without smell to inform me of what he'd found in the garbage, I allowed him to lick my mouth with his warm, wet tongue. Mel's delight in his ability to distract me from my work with ball-tossing—his delight in me—began to rub off. How had I earned such avid affection and complete trust?

Our afternoon walks were the best part of my day. I'd tantalize Mel with a slow preparation waltz around the kitchen, first to the drawer with the plastic zip-lock bags, just in case, then to the cabinet where we kept the pet gear—heartworm medicine, Sweetie's papers, and the purple velvet pouch that concealed the cold metal urn with her ashes inside it, destined for a spring burial in the garden. I'd pretend to debate which leash we'd use that day—this was Mel's cue to begin his mincing two-step, performed entirely on his hind legs, ears flopping as he yipped in excitement, *All right, already! Let's go!*

Mel had me well trained. I followed him to the door and outside. Sometimes when no one seemed to be around (if it was well below freezing and there was a howling wind) I let him off the leash and watched him dart across the neighbors' yards in his Z formation, zipping this way after a squirrel until he forgot what he was after, then trotting a few steps before zipping that way to chase a rabbit. A perpetual sniffing machine, he mostly just tore along with his nose to the ground, circling trees and fire hydrants and marking them. Sometimes he'd stand perfectly still and lift his nose in the air and maybe one of his front paws too. Mel did a great impression of a Labrador pointing.

Mel was my olfactory surrogate. I loved pretending that I could smell the same galvanizing odors he could—which hadn't been true even when I had a working nose. So why *not* pretend, just as I'd pretended to be a cowgirl when I was a kid, careening around the yard with a broom between my legs? Why couldn't I be a dog instead of a middle-aged anosmic? Something of my mother's capacity for turning the clock back must have been handed down to me.

At the same time, my worst fear had not come true. I was not losing touch with reality. Even though I had lost my sense of smell, or maybe because of that, I was looking inward with fresh eyes, and I could feel a different door to the outside world opening, albeit on squeaky hinges.

One sunny day Mel and I were out walking and took a different route, passing by a popcorn shop. I hadn't been aware of the shop until something abruptly switched the channel in my brain from one station (the dahlia network) to another. The topic that came so out of the blue (not to mention out of sync with the calendar) had visuals and a soundtrack. There was Cam in the kitchen watching football, a huge bowl of popcorn by his side, yelling at the players whose careers he followed as if each and every one of them were his own firstborn son. I lifted my eyes from the pavement just in time to miss colliding with a young woman leaving the shop with a full bag of popcorn. I had not seen the young woman or even the popcorn shop. I'd . . . only . . . smelled . . . the . . . popcorn.

Curiously enough, popcorn is one smell that bores Mel. He was straining so hard at the leash he was making himself gag. The aroma vanished and I swept all thoughts of popcorn into a mental dumpster that I kept firmly lidded. It also contained dangerous smell-related emotions such as hope, disappointment, and anger. They were all shameful. Why couldn't I stop thinking about the smells I used to love? Now I'd begun inventing them. I *was* like the man in Sacks's story. Pretty soon I'd be smoking a pipe too, and "snuffing and sniffing the spring" just the way that poor deluded fellow did. Mel gave me his conspiratorial look, the one that intimated we were partners in crime and it was time to make our move on the bank teller's window.

Mel's presence in my life was all the more wonderful for its improbability. He was like the stray violets, poppies, and columbines, those cheerful self-seeders that we gardeners affectionately refer to as volunteers. Mel had come out of left field. The best things always do. I remembered how the love for this yappy little dog that I'd held back for Sweetie's sake came on like a gusher the night she died. If I had to choose between Mel and smelling again, it would be a tough call.

He gave another hard tug on the leash. The awkward squat told me to pull out the plastic bag and make it quick. His over-the-shoulder glance seemed to register his acute embarrassment: *I know what you're thinking, but it only* looks *like I'm taking a crap in public; this is actually a restroom.* I plopped Mel's poop into the bag and sniffed in spite of myself. I could feel my upper lip curl in response to . . .

Stop this! I closed the zip-lock bag. The day had turned unseasonably hot and muggy, the air heavy with rain, I hoped. We needed it.

May would bring the lilacs. After a while I'd forget the old perfume. It would be as if they'd never had one. Weeks later, as I worked in my garden, the sun felt warm on my bare shoulders. The clammy soil between my fingers brought back the same fond memories it always had: of the previous spring, and the spring before that, and all the springs and summers of my childhood. I was thrilled by the colors of the flowers, even more by the subtle shades of green, gray, and gold that sometimes sprinkle, sometimes splotch, and sometimes cover completely the velvety or smooth or prickly surfaces of the leaves.

The catmint needed attention. If its flowers aren't snipped at just the right time, their seedlings pop up everywhere. I looked around for my hand pruners, found them. But something in the still air compelled me to set the pruners aside and bend closer to the coarse blue-gray foliage. Squeezing a leaf to release the scent, I slowly brought my fingers up to my nose.

Now I felt certain that my memory wasn't playing tricks. Licorice. With a hint of pine needles. Do we smell a touch of cumin? I'd never noticed it in a catmint leaf before. Or noticed the herb's essential . . . soapiness. Or the subtle aroma of wood smoke in the dry soil, and how the earth's smells "seem to multiply and extend" just before a thunderstorm, as Helen Keller wrote.

And wasn't it a coincidence — or the work of the gods, more likely — that just now clouds were gathering in the west? Sure enough, as "the storm draws nearer, my nostrils dilate, the better to receive the flood of earth-odors . . . until I feel the splash of rain against my cheek." This moment was just the way she'd described it.

I lay back on the grass, closed my eyes, opened my mouth, dilated my nostrils, and inhaled the fresh mineral flavors of the pouring rain. In a few minutes it was over. My skin rippled under a slight, exhilarating chill. "As the tempest departs, receding farther and farther, the odors fade, become fainter and fainter, and die away beyond the bar of space."

My journey was done. I was home again, and whole. The world was more intoxicating than ever before. I could really smell it.

Epilogue

THE STORY BROKE two days before my birthday: Zicam Cold Remedy nasal gel spray had finally come to the end of its long and profitable career. Calling the decision a warning to the drug industry that her agency was back in the business of protecting consumers' health, Margaret Hamburg, President Obama's new FDA commissioner, announced on June 16, 2009, that she'd be asking for an immediate recall of Zicam's intranasal zinc products. The company's other products, including oral zinc products, could still be sold. The FDA revealed that in addition to the known anosmia victims who'd participated in the class-action suit, there were more than one hundred others whose complaints to the agency had fallen on deaf ears. After the June 2009 announcement, the company first refused to even discuss the FDA recall with reporters, and then it vowed to fight the recall and placed full-page ads in newspapers across the country touting its oral cough suppressants, decongestants, throat-soothing agents, sinus-pain relievers, and vitamin-fortified immune-system boosters. (It still maintains that Zicam nasal products are safe and effective.) But the jig was up. Matrixx executives must have known they were lucky to get away with their company intact, their stock still trading (though after the FDA's announcement it immediately lost 70 percent of its value), and their pockets full. Their victims, meanwhile, the ones whose noses did not wake up, just have to keep on hoping that someday soon honest medical practitioners will find a cure.

Friends often ask me what it felt like, getting my smell back: Was it really just in that single moment with the catmint that you knew? It was. Undoubtedly my nose made steady improvement for months afterward. But I put an end to Cam's smell tests that day I smelled the catmint. I didn't want to tempt fate or appear to be greedy for more than what nature, in her infinite mercy, had restored to me already. I avoided any sort of deliberate, conscious exploration of my new olfactory reality.

But one sunny day in late May, not long after my sense of smell came back, I asked my husband if he'd like to walk to the park. He didn't have to ask which park, though there are several in the neighborhood; he knew I meant the one just two doors down from the house he grew up in, and just two blocks away from where we lived now. This nameless park is small, and secret, a vacant lot enclosed by lilacs. In the 1950s, Cam's parents and their neighbors had pooled their resources to buy the lot from the city. They wanted play space for their kids. All these years later, the lilacs still shade the straight dirt paths of an abandoned baseball diamond (kids go to Little League fields these days). They still send their heavy scent into the adjacent houses, which used to tower over them but which they can now look straight in the eye.

We stood on the sidewalk and smelled the lilacs. I felt no need to shove my nose into the flowers themselves, but Cam wanted to celebrate. He'd brought along my Felco pruners. Now he snipped a few especially plump and fragrant blooms, made a bouquet, and presented it to me with tears in his eyes. I obliged him by burying my nose in the flowers until the smell began to fade—adaptation, of course—and I felt a twinge of panic.

We brought the flowers home. I worried as we walked that I'd overdone it, the way you can after a bout of flu and end up back in bed. Cam hurried into the house with the flowers to find a vase. I lingered outside, trying not to think about smell, knowing that thinking would only drive it farther away. By the time I went inside and into the kitchen, I'd almost forgotten the lilacs. Their scent bowled me over even before I saw them. They were sitting in the center of the kitchen table in a beautiful glass vase. My nose had passed its first real test. From then on I knew we were home free.

That winter Cam and I drove to Chicago to celebrate (again) the return of my sense of smell, this time with an eleven-course meal at Alinea. Grant Achatz was doing well. His business partner Nick Kokonus said the recession has been tough on the restaurant—it used to take months to get a table, but now people had to wait only a few weeks to get in. The restaurant was full when Cam and I had our dinner to honor my nose. Grant invited us out to the kitchen to say hello. Slender and pale, with straight, shoulder-length brown hair, he can only be described as elfin in appearance. He thanked us for coming. His tongue has healed. He talks without difficulty and tastes as well as before. He says he feels fine.

Becky Phillips hasn't stopped hoping that her nose will recover. She couldn't stop hoping and still go on living happily, she said. Just as I'd embarked on a quest for answers through science, she'd set about deconstructing the old Becky, the queen of ambiance. She believed that by reactivating the still-intact (not anosmia-destroyed) portion of this person, she might retrieve a semblance of her once cheerful and independent self. After careful thought, she concluded that the still-intact piece was her love of ritual. Planning a party. Preparing food. Drawing a bath. Knowing that for an hour or more her time and thoughts were booked. Nothing else mattered but what she was doing then. Fragrance had been a way of getting to a place where time stopped and all the things she worried about just vanished.

"When I cook now, I slow way down. I make a ritual out of the steps and focus on each one of them. I enjoy the shopping, selecting ingredients. I bought a set of very fancy and beautiful knives that make chopping and slicing a tactile and a visual pleasure. I take a lot of time setting the table, arranging the food on the plate. I found that my taste buds are good for something. I like Indian and Mexican dishes and I season everything way more than I used to."

She reminds herself of the tradeoffs, and her belief that the good outweighs the bad. At the top of the good list is her husband. Joe lost his first wife, who was Becky's best friend, to a stroke the same year Phillips was hurt. The three had been close since childhood. Becky and Joe were drawn to each other by the enormous loneliness they were each coping with, and by the affection that had taken root in empathy, a shared appreciation for what the other had lost: Becky a cherished part of herself and Joe his spouse, who also happened to have been Becky's closest friend. "We both needed healing. We healed each other."

She closed our conversation by telling me she had shopping to do. She was going on a diet. This meant restocking the refrigerator. Ritual, she said, will help with the "satiety" problem. She's developing an act five. Snuff out the candles and clear the dishes. Arrange them neatly on the counter. Fill the sink with the detergent that used to delight her with its fragrance. Focus on its texture instead and on the warm suds. Remove the remains of dinner from each lovely china plate, towel it dry, put it away in a clean, orderly cupboard.

When she's done with the dishes, she will leave the kitchen (instead of hanging around watching TV and snacking on leftovers), turn out the light, and join her husband in the living room. If she's lucky he will have put on some music and built a fire.

Acknowledgments

I was an editor for years before becoming a writer. Deanne Urmy, who bought this book back in 2006 after reading my brief proposal, taught me what brilliant editors actually do. The experience has been humbling, to say the least. Deanne kept the faith as I balked once it dawned on me what I'd gotten myself into and tried to sell her a memoir instead of the science book she'd commissioned. She helped me find the book's elusive structure and protected my credibility as a narrator who fell back on "the yucks," as she put it, when I didn't believe in the importance of my own story. She is a perfectionist who is passing on her habits to future editors, like her assistant Nicole Angeloro, who also helped me enormously. Even in this time when books are declared an endangered species, tenacious editors like Deanne who live and die by excellence in literature are making sure that the written word will never fall out of favor, become irrelevant, or go extinct.

Deanne sent the book to Tracy Roe to copyedit. Tracy is a physician and a copyeditor who lives in Richmond, Virginia. *Breathtaking* is the only word I can come up with (Tracy would have a better one) to describe her consistently spot-on suggestions, not to mention the compassion expressed in happy faces she sprinkled through the manuscript. She was as concerned about my feelings as she was about getting the book right, down to the last semicolon.

My husband and daughters, as well as my siblings, kept my spirits up as I worked my way through the anosmia and then through the book I felt compelled to write about it. My sister, Judy Titcomb, read various drafts and offered excellent advice, as did my brother Bruce Leslie. My brother Frank Leslie always knew when to ask how the book was coming and when not to bring it up.

I'd also like to thank the many scientists and science writers who made this book possible. Richard Doty, Matt Ridley, Yilad Goav, Gordon Shepherd, Don Wilson, Chandler Burr, and Jonah Lehrer responded to the pleading e-mails of a writer who didn't know a gene from a cell (or a cell from a dahlia tuber, for that matter) when she embarked on what turned out to be a ludicrously ambitious project. The warm welcome I received at Richard Axel's lab at Columbia University is described in the book. Many, many thanks for a most memorable morning, Dan.

Two longtime writer friends also lent a hand. William Swanson, author of *Dial M: The Murder of Carol Thompson*, and William Souder, whose most recent book is *Under a Wild Sky* and whose next book is a biography of Rachel Carson, both made thoughtful suggestions that gave *Remembering Smell* a new lease on life when I was at my wits' end.

Bibliography

Achen, Alexandra C., and Frank P. Stafford. "Data Quality of Housework Hours in the Panel Study of Income Dynamics: Who Really Does the Dishes?" Institute for Social Research, University of Michigan at Ann Arbor, September 2005.

Ackerman, Diane. *A Natural History of Love.* New York: Random House, 1994.

———. *A Natural History of the Senses.* New York: Random House, 1990.

Aftel, Mandy. *Essence and Alchemy: A Natural History of Perfume.* New York: North Point Press, 2001.

Ashkenazy, Daniella. "A Mothering Touch." *Israel Magazine-on-Web,* January 1, 2001.

Axel, Richard. "Scents and Sensibility: Towards a Molecular Logic of Perception." Paper presented at the Columbia University Symposium on Brain and Mind, Columbia University, New York City, May 13, 2004.

Axel, Richard, and Linda Buck. "A Novel Multigene Family May Encode Odorant Receptors: A Molecular Basis for Odor Recognition." *Cell* 65 (1991): 175–87.

Axel, Richard, et al. "Allelic Inactivation Regulates Olfactory Receptor Gene Expression." *Cell* 78 (2004): 823–34.

Bakalar, Nicholas. "Childhood: Mold and Pollen May Affect Asthma Risk." *New York Times,* March 2, 2009.

Basu, Paroma. "Marmoset Dads Don't Stray." *University of Wisconsin–Madison News,* March 16, 2005.

Bayard, Tania, et al. *Gardening for Fragrance*. Brooklyn: Brooklyn Botanic Garden, 1989.

Blanke, Olaf, and Shahar Arzy. "The Out-of-Body Experience: Disturbed Self-Processing at the Temporo-Parietal Junction." *Neuroscientist* 11 (2005): 16–24.

Boodman, Sandra. "The Men Behind Zicam." *Washington Post,* January 31, 2006.

———. "Paying Through the Nose: Maker of Cold Spray Settles Lawsuits for $12 Million but Denies Claim that Zinc Product Ruined Users' Sense of Smell." *Washington Post,* January 31, 2006.

Borm, Paul J. A., et al. "Exposure to Diesel Exhaust Induces Changes in EEG in Human Volunteers." *Particle and Fibre Toxicology* 5 (2008): 4.

Brumfield, C. Russell. *Whiff! The Revolution of Scent Communication in the Information Age*. New York: Quimby Press, 2008.

Buck, Linda, and Stephen Liberles. "A Second Class of Chemosensory Receptors in the Olfactory Epithelium." *Nature* 442 (2006): 645–50.

Burr, Chandler. *The Emperor of Scent*. New York: Random House, 2002.

Burstein, A. "Olfactory Hallucinations." *Hospital & Community Psychiatry* 38 (1987): 80.

Butler, Chris, et al. "Transient Epileptic Amnesia: Regional Brain Atrophy and Its Relationship to Memory Deficits." *Brain* 132 (2009): 357–68.

Byron, Ellen. "Is the Smell of Moroccan Bazaar Too Edgy for American Homes?" *Wall Street Journal,* February 3, 2009.

Calderón-Garcidueñas, L., et al. "Air Pollution, Cognitive Deficits and Brain Abnormalities: A Pilot Study with Children and Dogs." *Brain and Cognition* 68 (2008): 117–27.

Callaway, Ewen. "Neanderthal Genome Already Giving Up Its Secrets." *New Scientist,* December 10, 2008.

Carlson, J., et al. "Odor Coding in the Maxillary Palp of the Malaria Vector Mosquito *Anopheles gambiae*." *Current Biology* 17 (2009): 1533–44.

Chorost, Michael. *Rebuilt: How Becoming Part Computer Made Me More Human*. Boston: Houghton Mifflin, 2005.

Chu, Simon, and John J. Downes. "Odour-Evoked Autobiographical Memories: Psychological Investigations of Proustian Phenomena." *Chemical Senses* 25 (2000): 111–16.

Classen, Constance, David Howes, and Anthony Synnott. *Aroma: The Cultural History of Smell*. London: Routledge, 1994.

Corbin, Alain. *The Foul and the Fragrant: Odor and the French Social Imagination.* Cambridge, MA: Harvard University Press, 1986.

Cytowic, Richard. *The Man Who Tasted Shapes.* New York: G. P. Putnam's Sons, 1993.

Damasio, Antonio. *Descartes' Error: Emotion, Reason, and the Human Brain.* New York: G. P. Putnam's Sons, 1994.

Dawkins, Richard. *The Selfish Gene.* Oxford: Oxford University Press, 1989.

DePass, Dee. "For Aveda, the Problem with Common Scents." *Minneapolis–St. Paul Star Tribune,* August 17, 2008.

Doty, Richard L. *The Great Pheromone Myth.* Baltimore: Johns Hopkins University Press, 2009.

————. *The Handbook of Olfaction and Gustation.* 2nd ed. New York: Marcel Dekker, 2003.

Dulac, Catherine, Tali Kimchi, and Jennings Xu. "A Functional Circuit Underlying Male Sexual Behaviour in the Female Mouse Brain." *Nature* 448 (2007): 1009–14.

Dulac, Catherine, et al. "Olfactory Inputs to Hypothalamic Neurons Controlling Reproduction and Fertility." *Cell* 123 (2005): 669–82.

Eidelman, A. I., et al. "Mothers' Recognition of Their Newborns by Olfactory Cues." *Developmental Psychobiology* 20 (1987): 587–91.

Elder, Alison, et al. "Translocation of Inhaled Ultrafine Manganese Oxide Particles to the Central Nervous System." *Environmental Health Perspectives* 114 (2006): 1172–78.

Engen, Trygg. *Odor Sensation and Memory.* New York: Praeger Publishers, 1991.

Felten, Eric. "These Drams Are Different." *Wall Street Journal,* March 18, 2006.

Firestein, Stuart, and J. M. Otaki. "Length Analyses of Mammalian G-protein-coupled Receptors." *Journal of Theoretical Biology* 211 (2001): 77–100.

Firestein, Stuart, et al. "The Molecular Receptive Range of an Odorant Receptor." *Nature Neuroscience* 3 (2000): 1248–55.

Frasnelli, J., et al. "Clinical Presentation of Qualitative Olfactory Dysfunction." *European Archives of Oto-Rhino-Laryngology* 261 (2004): 411–15.

Frasnelli, J., and T. Hummel. "Olfactory Dysfunction and Daily Life." *European Archives of Oto-Rhino-Laryngology* 262 (2005): 231–35.

Gawande, Atul. "The Itch." *The New Yorker,* June 30, 2008.

Geddes, Linda. "New Kind of Epilepsy Shakes Up Memory." *New Scientist,* February 2, 2009.

Gee, Mike. *The Final Days of Michael Hutchence.* London: Omnibus Press, 1998.

Geffen, Maria Neimark, et al. "Neural Encoding of Rapidly Fluctuating Odors." *Neuron* 61 (2009): 570–86.

Gerstein, Mark, and Deyou Zheng. "The Real Life of Pseudogenes." *Scientific American,* August 2006.

Gilad, Yoav, et al. "Loss of Olfactory Receptor Genes Coincides with the Acquisition of Full Trichromatic Vision in Primates." *Public Library of Science Biology* 2, no. 1 (January 2004). http://www.ncbi.nlm.nih.gov/pmc/articles/PMC314465/.

Gilbert, Avery. *What the Nose Knows: The Science of Scent in Everyday Life.* New York: Crown, 2008.

Gilbert, Daniel. *Stumbling on Happiness.* New York: Alfred A. Knopf, 2006.

Groopman, Jerome. "That Buzzing Sound." *The New Yorker,* February 9, 2009.

Harkema, Jack, Zahidul Islam, and James Pestka. "Satratoxin G from the Black Mold *Stachybotrys chartarum* Evokes Olfactory Sensory Neuron Loss and Inflammation in the Murine Nose and Brain." *Environmental Health Perspectives* 114 (2006): 1099–1107.

Harvey, Susan Ashbrook. *Scenting Salvation: Ancient Christianity and the Olfactory Imagination.* Berkeley: University of California Press, 2006.

Herz, Rachel. "A Naturalistic Analysis of Autobiographical Memories Triggered by Olfactory Visual and Auditory Stimuli." *Chemical Senses* 29 (2004): 217–24.

———. *The Scent of Desire: Rediscovering Our Enigmatic Sense of Smell.* New York: HarperPerennial, 2007.

Hirsch, Alan. *Scentsational Sex.* Boston: Element Books, 1998.

———. *What Flavor Is Your Personality?* Chicago: Sourcebooks Inc., 2001.

Hirsch, Alan, and M. L. Colavincenzo. "An Investigation into the Effects of Certain Odors Upon Anxiety." *Journal of Neurological and Orthopaedic Medicine and Surgery* 22 (2002): 47–53.

Hull, John M. *Touching the Rock: An Experience of Blindness.* New York: Pantheon Books, 1990.

Hummel, Thomas, and Antje Welge-Lüssen, eds. *Taste and Smell: An Up-*

date. Advances in Oto-Rhino-Laryngology 63. Basel, Switzerland: Karger, 2006.

Igarashi, Kei M., and Kensaku Mori. "Spatial Representation of Hydrocarbon Odorants in the Ventrolateral Zones of the Rat Olfactory Bulb." *Journal of Neurophysiology* 93 (2005): 1007–19.

Insel, Thomas R., et al. "A Role for Central Vasopressin in Pair Bonding in Monogamous Prairie Voles." *Nature* 365 (1993): 545–48.

Johnston, Amy, et al. "Olfactory Ability in the Healthy Population." *Chemical Senses* 31 (2006): 763–71.

Kaitz, M., and A. I. Eidelman. "Smell-Recognition of Newborns by Women Who Are Not Mothers." *Chemical Senses* 17 (1992): 225–29.

Keller, Helen. *The Story of My Life.* New York: Signet,1988.

Kessler, David A. *The End of Overeating.* Emmaus, PA: Rodale, 2009.

Kurson, Robert. *Crashing Through: A True Story of Risk, Adventure, and the Man Who Dared to See.* New York: Random House, 2007.

LeDoux, Joseph. *Synaptic Self: How Our Brains Become Who We Are.* New York: Viking Penguin, 2002.

Lee, Thomas. "Local Firms Hope to Spark Solution for Obesity." *Minneapolis–St. Paul Star Tribune,* March 3, 2009.

Lehrer, Jonah. *How We Decide.* Boston: Houghton Mifflin Harcourt, 2009.

———. "The Migration of New Neurons." The Frontal Cortex, February 16, 2009. http://scienceblogs.com/cortex/.

———. *Proust Was a Neuroscientist.* Boston: Houghton Mifflin, 2007.

Leopold, Donald. "Distortion of Olfactory Perception: Diagnosis and Treatment." *Chemical Senses* 27 (2002): 611–15.

Lin, Da Yu, Stephen Shea, and Lawrence Katz. "Representation of Natural Stimuli in the Rodent Main Olfactory Bulb." *Neuron* 50 (2006): 937–49.

Linden, David J. *The Accidental Mind: How Brain Evolution Has Given Us Love, Memory, Dreams, and God.* Cambridge, MA: Belknap Press of Harvard University Press, 2007.

MacLean, Paul D. *The Triune Brain in Evolution: Role in Paleocerebral Functions.* New York: Plenum Press, 1989.

Masson, Jeffrey Moussaieff. *Assault on Truth: Freud's Suppression of the Seduction Theory.* New York: Farrar, Straus and Giroux, 1984.

Max, D. T. "A Man of Taste." *The New Yorker,* May 12, 2008.

McClintock, Martha. "Menstrual Synchrony and Suppression." *Nature* 229 (1971): 244–45.

McClintock, Martha, et al. "Psychological Effects of Musky Compounds: Comparison of Androstadienone with Androstenol and Muscone." *Hormones and Behavior* 42 (2002): 274–83.

Miller, Jeff. "Drosophila Envy: A Maverick Scientist Is Convinced Fruit Flies Have Much to Teach About the Evolution and Neurobiology of Emotions." *Howard Hughes Medical Institute Bulletin,* November 2006.

Mombaerts, Peter. "Love at First Smell—The 2004 Nobel Prize in Physiology or Medicine."*New England Journal of Medicine* 351 (2004): 2579–80.

Mombaerts, Peter, et al. "Structure and Emergence of Specific Olfactory Glomeruli in the Mouse." *Journal of Neuroscience* 21 (2001): 9713–23.

Morgenson, Gretchen. "The Eyeshade Smelled Trouble." *New York Times,* June 27, 2009.

Myers, David G. *Psychology.* New York: Worth Publishers, 2003.

Nagourney, Eric. "Screening: Higher Rates of Hearing Loss Are Found." *New York Times,* April 5, 2008.

Ober, Carole, Dagan Loisel, and Yoav Gilad. "Sex-Specific Genetic Architecture of Human Disease." *Nature Reviews Genetics* 9 (2008): 911–22.

Pääbo, Svante, et al. "A Complete Neandertal Mitochondrial Genome Sequence Determined by High-Throughput Sequencing." *Cell* 134 (2008): 416–26.

Pasquina, Paul F., et al. "Mirror Therapy for Phantom Limb Pain." *New England Journal of Medicine* 357 (2007): 2206–07.

Pearce, Jeremy. "Paul MacLean, 94, Neuroscientist Who Devised Triune Brain Theory, Dies." *New York Times,* January 10, 2008.

Pines, Maya. "The Mystery of Smell: The Memory of Smells." Report from the Howard Hughes Medical Institute, 2008.

Quammen, David. "Darwin's First Clues." *National Geographic,* February 2009.

Ramachandran, V. S., and W. Hirstein. "The Perception of Phantom Limbs. The D. O. Hebb Lecture." *Brain* 121 (1998): 1603–30.

Rayner, Richard. "Bug Wars." *The New Yorker,* August 25, 2008.

Ridley, Matt. *Francis Crick: Discoverer of the Genetic Code.* New York: Harper Collins, 2006.

———. "Modern Darwins." *National Geographic,* February 2009.

———. *Nature via Nurture: Genes, Experience, and What Makes Us Human.* London: Guardian Books, 2003.

Romano, Jay. "The Dangers of Mold in Homes." *New York Times,* October 7, 2001.

Rose, Steven. *Lifelines: Life Beyond the Gene.* London: Allen Lane, 1997.

Rosenblum, Lawrence. "You Drink What You Think." *Psychology Today,* August 23, 2009.

Sacks, Oliver. *A Leg to Stand On.* New York: Touchstone, 1984.

———. *The Man Who Mistook His Wife for a Hat.* New York: Touchstone, 1970.

Sanders, Lisa. "Fear of Falling." *New York Times Magazine,* September 6, 2009.

Shepherd, Gordon. "Smell Images and the Flavour System in the Human Brain." *Nature* 444 (2006): 316–21.

———. "A Theory of Olfactory Processing and Its Relevance to Humans." *Chemical Senses* 30 (2005): 3–5.

Shepherd, Gordon, and Kristen Shepherd-Barr. "Madeleines and Neuro-modernism: Reassessing Mechanisms of Autobiographical Memory in Proust." *Auto/Biography Studies* 13 (1998): 39–60.

Shubin, Neil. *Your Inner Fish: A Journey into the 3.5-Billion-Year History of the Human Body.* New York: Pantheon, 2008.

Singer, Natasha. "An Underdog Pursues the Scent." *New York Times,* August 20, 2008.

Sobel, Noam, et al. "Brain Mechanisms for Extracting Spatial Information from Smell." *Neuron* 47 (2005): 581–92.

Sobel, Noam, et al. "Mechanisms of Scent-Tracking in Humans." *Nature Neuroscience* 10 (2007): 27–29.

Spehr, M., et al. "One Neuron—Multiple Receptors: Increased Complexity in Olfactory Coding?" *Science Signaling* 285 (2005): 25.

Stanford, Matthew. *Waking: A Memoir of Trauma and Transcendence.* Emmaus, PA: Rodale Press, 2006.

Stoddart, D. Michael. *The Scented Ape: The Biology and Culture of Human Odour.* Cambridge: Cambridge University Press, 1990.

Stuefer, Josef. "Clever Plants Chat over Their Own Network." Netherlands Organization for Scientific Research, September 4, 2007.

Süskind, Patrick. *Perfume: The Story of a Murderer.* New York: Alfred A. Knopf, 1976.

Svoboda, Elizabeth. "Sniff Test May Signal Disorders' Early Stages." *New York Times,* August 14, 2007.

Swaminathan, Nikhil. "The Scent of a Man." *Scientific American,* September 2007.

Talan, Jamie. "Intranasal Delivery of Stem Cells Bypasses Blood-Brain Barrier." *Neurology Today* 9 (2009): 1, 11–13.

Taylor, Jill Bolte. *My Stroke of Insight.* New York: Viking, 2008.

Turin, Luca. *The Secret of Scent: Adventures in Perfume and the Science of Smell.* New York: HarperCollins, 2006.

Vida, Vendela. "Scents and Sensibility: Has American Fiction Been Deodorized?" Slate.com, October 2004.

Vogt, Richard G. "How Sensitive Is a Nose?" *Science Signaling* 322 (2006): 8.

Vroon, Piet. *Smell: The Secret Seducer.* New York: Farrar, Straus and Giroux, 1997.

Wakin, Daniel J. "Verdi with Popcorn, and Trepidation." *New York Times,* February 15, 2009.

Wedekind, Claus, and Dustin Penn. "MHC Genes, Body Odours, and Odour Preferences." *Nephrology Dialysis Transplantation* 15 (2000): 1269–71.

Wedekind, Claus, et al. "MHC-Dependent Mate Preferences in Humans." *Proceedings: Biological Sciences* 260 (1995): 245–49.

Wilson, Donald A., and Richard J. Stevenson. *Learning to Smell: Olfactory Perception from Neurobiology to Behavior.* Baltimore: Johns Hopkins University Press, 2006.

Wilson, Edward O. *On Human Nature.* Cambridge, MA: Harvard University Press, 1978.

Wyart, Claire, et al. "Smelling a Single Component of Male Sweat Alters Levels of Cortisol in Women." *Journal of Neuroscience* 27 (2007): 1261–65.

Ziegler, Toni. "Neuroendocrine Response to Female Ovulatory Odors Depends upon Social Condition in Male Common Marmosets." *Hormones and Behavior* 47 (2005): 56–64.

Ziegler, Toni, et al. "Exposure to Infant Scent Lowers Serum Testosterone in Father Common Marmosets (*Callithrix jacchus*)." *Biology Letters,* December 23, 2008; doi:10.1098/rsbl.2008.0358.

Zierah, Elizabeth. "The Nose That Never Knows: The Miseries of Losing One's Sense of Smell." Slate.com, July 8, 2008.

Notes

page 1. A POWERFUL STENCH

12 *burning mouth syndrome:* Grushka and Ching, "Burning Mouth Syndrome," in Hummel and Welge-Lüssen, eds., *Taste and Smell,* 278–87. Symptoms are extreme burning sensations in the mouth as well as changes in the taste and other sensory systems. Secondary causes may range from a drug allergy to a low thyroid condition to psychosomnia.

2. DIAGNOSIS

15 *Zicam:* The Zicam Cold Remedy recall was announced by the FDA on June 16, 2009. The first such recall under the Obama administration, it was widely viewed as signaling a shift in regulatory policy toward more consumer protection.

5. AN UNDERLYING LOGIC

40 *Yoav Gilad sparked:* Gilad et al., "Loss of Olfactory Receptor Genes." Evolution has apparently been disabling olfactory genes in exchange for improvement in other senses; findings in certain primates suggest that the deterioration of their olfactory ability occurred at the same time as their acquisition of color vision. Superior vision in monkeys, as in humans, may have fostered left-brain language skills — also at the expense of smell.

41 *The prevailing theory:* Burr, *Emperor of Scent,* 29, 30.

42 *glomeruli:* Lin, Shea, and Katz, "Representation of Natural Stimuli." Utilizing gas chromatography to identify specific molecules in an odor compound sniffed by a rat and then using intrinsic signal imaging to map brain activity in response, researchers at Duke showed that each glomerulus responded to one specific odorant only. They hypothesized that "there are no single detectors for complete smells." In other words, the integration of all the individual stimuli is done in more advanced brain structures (the olfactory cortex).

Peter Mombaerts: Mombaerts, "Love at First Smell." Peter Mombaerts was a postdoc in the Axel lab for many years and is now at Rockefeller University. See also Mombaerts et al., "Structure and Emergence."

6. THE BREAKTHROUGH

45 *Linda Buck:* Shubin, *Your Inner Fish*, 143–44. Neil Shubin describes the assumptions that led to the smell-gene discovery.

46 *the Austrian philosopher:* Rose, *Lifelines*. In *Lifelines: Life Beyond the Gene*, biologist Steven Rose devotes a chapter to a critique of reductionism (reductionism is the attempt to explain complex systems by reducing them to their smaller, simpler components). Karl Popper, whose system of proving or disproving hypotheses (the scientific method) made him an early proponent of reductionism, was among those who grew to see that when the principle was rigidly applied it was too restrictive. The philosopher Thomas Kuhn, perhaps the most influential of the reductionist reformers, believed that scientists should develop paradigms — overarching theories — of how the world works. This approach is both objective and intuitive. Moreover, new technologies, such as MRI scanners, that allow scientists to observe genetic influences in living creatures have profoundly changed the nature of experimentation itself.

Buck found the genes: Burr, *Emperor of Scent*, 34.

48 *Axel jokingly calls:* Axel, "Scents and Sensibility." Axel elaborated on this, saying that how any collection of sensory inputs "ultimately elicits appropriate behavioral or cognitive responses is what Vernon Mountcastle has described as 'the big in-between, the ghost in the machine.'"

Scientists have shown: Groopman, "That Buzzing Sound."

7. PHANTOMS

49 *had me wondering:* Frasnelli et al., "Qualitative Olfactory Dysfunction."

51 *A California psychologist:* Herz, *Scent of Desire,* 77–78.
    amputees who come home: See Pasquina et al., "Mirror Therapy for Phantom Limb Pain."

52 *Oliver Sacks described:* Sacks, *Leg to Stand On,* 141.

54 *Experiments done in European labs:* Blanke and Arzy, "Out-of-Body Experience." Blanke found that he could also "plant" physical sensations in the subjects' extremities at will. The work was done at the Ecole Polytechnique in Lausanne, Switzerland; similar experiments have been conducted at University College London and Princeton University.

55 *a woman who had:* Gawande, "Itch."

57 *I decided to find out:* This section is based on e-mails I exchanged with Frasnelli and neurobiologist Don Wilson in June 2007. The ideas that Wilson attributes to Joseph LeDoux are discussed in LeDoux's *Synaptic Self,* especially in chapter 7, "The Mental Trilogy."

8. THE SCENTLESS DESCENT

64 *Pain and suffering:* In "Olfactory Dysfunction and Daily Life," Johannes Frasnelli writes that in the few studies that have been done on smell loss and depression, it appears that depression (and related problems) are much more prevalent in people with anosmia and parosmia than they are in the general population. Olfactory improvement almost always parallels a significantly improved satisfaction with life.

    Becky Phillips lost: This is not her real name; it was changed to protect her privacy.

67 *Rachel Herz tells:* *Scent of Desire,* 1–3, 5, 16–17. Michael Hutchence was riding a bicycle in Copenhagen when he was hit by a car. Herz notes that he lost only his sense of smell but that he thought he'd lost both smell and taste; the experience of eating is so intertwined with smell that it's led to the common misconception that taste and flavor are the same thing. In fact, flavor is the combination of taste and smell. Herz says that depression after anosmia is almost

always progressive. Those who suddenly go blind are more traumatized in the immediate aftermath but improve over time, while
"follow-up analyses on the emotional health of [anosmics] one
year later showed that the anosmics were faring much more poorly
than the blind." See also Mike Gee's *Final Days of Michael Hutchence.*

69 *an idiot savant:* Engen, *Odor Sensation,* 3–5.

70 *Elizabeth Zierah had:* "Nose That Never Knows," Slate.com.

9. TASTING THE HOLIDAYS

74 *Rachel Herz had:* Herz, *Scent of Desire,* 197–98.
    Taste is said to: Estimates range from 80 to 90 percent, though it's
    impossible to measure this.
    One olfactory expert: Gilbert, *What the Nose Knows,* 91–102.

76 *cultural conditioning:* See Corbin, *Foul and the Fragrant.* How odor
    perception is formed by culture is also the subject of Constance
    Classen's *Aroma* and a prominent theme of Patrick Süskind's *Per-
    fume.*

78 *Most humans:* Callaway, "Neanderthal Genome." The *New Scientist*
    reported this early finding by study leader Svante Pääbo of the Max
    Planck Institute for Evolutionary Anthropology in Leipzig, Ger-
    many. Researchers will use the information to help determine pre-
    cisely when humans and Neanderthals took separate evolutionary
    paths and if there was interbreeding. See also Pääbo, "A Complete
    Neandertal Mitochondrial Genome."

79 *Autism makes it difficult:* Wilson and Stevenson, *Learning to Smell,* 21.
    The authors cite several syndromes that adversely affect odor mem-
    ory: "That we find relationships between stimulus and discrimina-
    tive ability and odor quality is not surprising, because these will of-
    ten have similar patterns of receptor activity."

10. IDIOT SAVANT

90 *Smell scientist Stuart Firestein:* Firestein is a geneticist at Columbia
    University and a leading expert on signal transduction. On his web-
    site he explains that "the olfactory neuron is uniquely suited for these
    studies since it is designed specifically for the detection and discrim-
    ination of a wide variety of small organic molecules, i.e., odors." His

studies of mammalian receptors have been published in *Nature Neu-roscience* and the *Journal of Theoretical Biology,* among other journals. *a recent contest:* Sobel, "Mechanisms of Scent-Tracking." Psychologist Noam Sobel conducted the contest. The humans proved that our species has an excellent scent-detecting system, thanks to an individual's twin nostrils. While dogs and humans both cycle back and forth between nostrils, only people can connect a smell to a conscious event or label. As Richard Doty pointed out, verbal scores increase when smells are delivered to the left (language) side of the brain. Commenting on the contest, Shepherd noted that "olfactory genes do not map directly onto smell acuity; dogs, which have superior tracking abilities, have only about 850 functional genes. In fact, behavioral tests show that primates have surprisingly good senses of smell, and it has been argued that the decline in olfactory-gene number is more than offset in humans by their much enlarged brains [capable of] analysis and complex processing of smell to guide critical human behaviors." Shepherd also suggested "that evolution has produced in humans an excellent overall sense of smell and, combined with taste and somatosensation and other inputs, the best sense of flavor in the animal world."

women score higher: Smell dysfunction expert Richard Doty told me that women usually outperform men on smell-identification tests.

II. SENSELESS EATING

98 *Kessler thinks food companies:* Kessler, *End of Overeating.*
100 *appetite is erased:* Linden, *Accidental Mind,* 15. In his book, Linden discusses new research on hunger and satiety.
Venture capital is pouring: Lee, "Local Firms."

12. CULINARY ART

103 *Born in Provence:* Lehrer, *Proust Was a Neuroscientist,* chapter 3. I am indebted to Jonah Lehrer for his overview of the evolution of fine cooking.
104 *Alinea challenges:* My husband and I dined at Alinea most recently in March 2009, and we met Grant Achatz, whose cancer was then in remission. Medical details of Achatz's cancer diagnosis and treatment are from Max, "Man of Taste."

13. OLFACTORY ART

108 *University of Michigan study:* Achen and Stafford, "Data Quality of Housework Hours."
    To entice both: Byron, "Smell of Moroccan Bazaar."

110 *the* New York Times: Chandler Burr was hired as perfume critic in 2006.

112 *chief perfumer at Aveda:* DePass, "For Aveda."
    Another innovation in perfumery: Singer, "Underdog Pursues."

14. A HISTORY OF THE SENSUAL NOSE

115 *the nineteenth-century Italian composer:* Wakin, "Verdi with Popcorn."

117 *Cultural anthropologist Constance Classen:* Classen, *Aroma,* 157–58.
    As flesh was despised: Harvey, *Scenting Salvation,* 129–30, 205.

118 *Patrick Süskind's* Perfume: Süskind, *Perfume,* 3, 14–15.

119 *between the "hillocks":* Ackerman, *Natural History of Love,* 237.
    cleaned up its act: Vida, "Scents and Sensibility."
    neurologist-turned-analyst Sigmund Freud: Details of the Fliess story were taken from the transcript of a 1986 interview with Jeffrey Moussaieff Masson on the ABC Radio program *The Science Show.* Also see Masson's *Assault on Truth.*

15. THE EROTIC NOSE

126 *males of almost:* Ridley, "Modern Darwins."
    A famous study: Wedekind et al., "MHC-Dependent Mate Preferences." See also Wedekind and Penn, "MHC Genes."

127 *Pherlure cologne:* This I found advertised online in 2006.

128 *in the animal world:* The prairie vole studies were conducted in the late nineties by Tom Insel and Sue Carter at Emory University's primate research center. Joseph LeDoux discussed their work in *Synaptic Self.* Insel believes that "the neural basis of attachment can be investigated in animal models." These neural pathways also may prove to be important in treating clinical disorders such as autism and schizophrenia, both of which result in social isolation and detachment.
    actually stick around: Basu, "Marmoset Dads." A senior scientist at the Wisconsin National Primate Research Center at UW–Madison, Toni Ziegler published her findings in *Biology Letters* in 2008. In

2006, Ziegler compared weight gain in both control and paternal marmosets and tamarins during the pregnancy and birth period.

129 *Catherine Dulac:* Catherine Dulac has identified MHC pathways in mice. She explains on her website: "The functional characterization of M10 highlights an unexpected role for MHC molecules in pheromone detection by mammalian VNO neurons and opens new avenues of research on the process of sensory detection leading to behavior." See Dulac, Kimchi, and Xu, "Functional Circuit," for the details of the mouse-pheromone experiment.

130 *human babies send:* Ashkenazy, "Mothering Touch." The article notes that Hebrew University psychologist Marsha Kaitz published data showing that mothers could identify their own children by smell 90 percent of the time. Subsequent research has shown that any woman who holds a baby long enough can pick it out from others by smelling it. See also Eidelman et al., "Mothers' Recognition."

An organ that: Axel called the VNO "the erotic nose" in answering a question after his 2004 speech to alumni at Columbia University.

Dulac and her colleagues: Dulac et al., "Olfactory Inputs." In 1995 Catherine Dulac and Richard Axel published work on gene-splicing techniques that enabled her to go on to focus on parsing the respective roles of olfaction and the VNO in mice. Back-to-back studies in 2005 confirmed that sexual activity is not limited to the VNO, which opens the door to the possibility that humans also have a pheromone-based system operating through smell.

131 *atrophies before birth:* This has been widely accepted for decades.

As Dulac explains: This quote from the Dulac lab website reads in full: "Our data contradict the established notion that VNO activity is required for the initiation of male-female mating behavior in the mouse and suggest instead a critical role in ensuring sex discrimination."

133 *or possibly the skin:* Herz, in *Scent of Desire*, page 145, speculates that "the chemicals responsible for inducing menstrual synchrony are transmitted directly through the skin. That is, the sweat of the donor female is absorbed through the skin of another woman through touch (sweat-to-skin contact), which over time enters her bloodstream, causing changes to her hormonal system that are in synch with the donor woman." It should be noted that, as Wilson

and Stevenson point out in *Learning to Smell* (page 139), "some of the reported synchrony effects may be artifacts of the recording procedure and that synchrony may not confer any benefits with respect to reproductive success." See also Doty, *Great Pheromone Myth.*

the Bruce effect: The widely studied phenomenon was referred to in Dawkins's *The Selfish Gene.*

134   *trees in Beijing:* Rayner, "Bug Wars."

Doty compares pheromones to Snarks: Doty makes the argument in *The Great Pheromone Myth* that "[w]ithout negating the fact that biological secretions ubiquitously influence mammalian social and sexual behaviors, as well as endocrine state, . . . in the vast majority of cases such influences cannot be divorced from the brain and are not fundamentally different from similar influences of other sensory systems."

normal sex lives: Doty's point is that people born without smell are "normal" in every other respect. They don't know what they're missing, that's true; but what they're missing, at least according to their own testimony, has no negative impact on their marriage, their interest in sex, their interest in life. They even enjoy food. My take on this is that they are no different from someone born lacking an arm or a leg, vision or hearing. The brain is plastic and begins adapting to such disabilities even in utero; what's traumatic is the sudden loss of a function that the brain is already wired to rely on.

135   *a religious sect:* Ober, Gilad, and Loisel, "Sex-Specific Genetic Architecture." University of Chicago geneticist Carole Ober has also done studies on the Hutterites with Martha McClintock.

136   *Swiss researchers:* Wedekind and Penn, "MHC Genes." In this study, men adhered to a strict bathing regimen and wore no deodorants or colognes.

the Berkeley study: Wyart et al., "Smelling a Single Component." A postdoctoral fellow working under Noam Sobel, an associate professor of psychology and director of the Berkeley Olfactory Research Program, Wyart hopes to harness her findings to help people with low cortisol levels, such as those patients with Addison's disease. Pills prescribed for Addison's can often cause ulcers and weight gain, but "[m]erely smelling synthesized or purified human chemosignals may modify endocrine balance."

137   *In a groundbreaking:* McClintock, "Menstrual Synchrony."

smell of clove: McClintock et al., "Psychological Effects."

16. THE LANGUAGE OF SMELL

140 *I came downstairs:* Felten, "These Drams Are Different."

141 *Andrei Codrescu:* "Life Without Smell May Not Be Worth It" aired on National Public Radio's *All Things Considered* on October 30, 2008.

146 *They do not:* MacLean, *Triune Brain,* 124–34.

A Dutch botanist: This work on underground runners and communication is being done by Josef Stuefer at Radboud University in the Netherlands.

147 *Neuroscientist Joseph LeDoux:* See LeDoux's *Synaptic Self,* 83–85, 259.

18. SMELL, MEMORY

160 *strength of a whiff:* Pines, "Mystery of Smell."

163 *Gottfried described:* Gottfried, "Assessment of Olfactory Function," in Hummel and Welge-Lüssen, eds., *Taste and Smell,* 98.

164 *When this region:* Shepherd and Shepherd-Barr, "Madeleines and Neuromodernism"; Rosenblum, "You Drink What You Think."

165 *Axel concluded:* Axel discussed Marcel Proust's *In Search of Lost Time* (a.k.a. *Remembrance of Things Past*), specifically volume 1, *Swann's Way,* in his Nobel acceptance speech, and in a speech at Columbia in 2004 he used a painting by Magritte with the phrase "Ce N'est Pas un Nez" superimposed over the artist's title to illustrate the confounding nature of sense perception.

Scientists and literature scholars: Chu and Downes, "Odour-Evoked Autobiographical Memories." These two professors at the University of Liverpool compiled and analyzed scientific studies of the Proust phenomenon going back to the very first one in 1928. They concluded that olfaction is unique in that it can summon memories that are older, more vivid, and more emotional than memories evoked by either verbal or visual cues, likely because of the smell-summoned memories' origin in the limbic system.

166 *Torrents of words:* In "Madeleines and Neuromodernism," the authors point out that the cookie-and-tea smell's familiarity was remarkable to Proust thanks to the intense euphoria it brought forth. The memory of Combray itself was far more difficult to unlock and came only through determined, conscious effort.

167 *olfaction's real champions:* Avery Gilbert plays devil's advocate in his book *What the Nose Knows.*

Jonah Lehrer disagrees: Lehrer, *Proust Was a Neuroscientist.* See chapter 4 for a full discussion. Memory breakthroughs are coming fast and furious these days. University of Edinburgh neuroscientist Chris Butler studies people with transient epileptic amnesia (TEA); these unfortunates have seizures that not only cause temporary amnesia but also erase selective memories in a seemingly random way. Butler's work challenges the notion that the hippocampus processes only short-term memories. His findings fit with a growing consensus that the ancient midbrain triumvirate—feeling, remembering, and smelling—drives human behavior more than thinking does. It also explains why people are able to "relive" past events and imagine the future. Finally, proving the ascendancy of the midbrain would settle the matter of which of the two—intuition or reductive reasoning—is the more reliable tool in decision-making. Lehrer discusses this in *How We Decide.* Noting that people who suffer midbrain neurological damage are often pathologically indecisive, Lehrer recommends following one's gut when long-term experience can be deployed in the decision-making process. As he explains it, dopamine neurons in the anterior cingulate cortex can warn of deviations from accustomed patterns. However, if one is faced with a novel situation, such as a nonpilot having to take the controls from a pilot who's suddenly incapacitated, better to put guesswork on hold and make radio contact with someone on the ground who can help guide the plane to a safe landing.

Spanish scientist: Lehrer, *Proust Was a Neuroscientist,* chapter 4.

19. A BARRIER BREACHED

175 *years and years:* Igarashi and Mori, "Spatial Representation." The two smell experts explained that a circuit in a perfumer's nose blends odorants to identify a smell mixture, such as banana. When the mixture is paired with a verbal or visual cue, a perfumer or taster can learn to recognize and identify highly complex aromas just by being around them a lot and by smelling them in conjunction with flashcard-like aids.

It used to be: Elder et al., "Translocation of Inhaled Ultrafine Manganese Oxide Particles."

children in California: Bakalar, "Childhood: Mold and Pollen." Researchers at UC–Berkeley studying some five hundred Hispanic children in California linked an increased risk of asthma for babies born in the winter months to those infants' exposure to mold. A 2008 study found that polluted air not only caused Alzheimer's lesions in dogs but also damaged children's brains. The first study to focus on nano particles was conducted in the Netherlands. Researchers found that even brief exposure (several hours) to the outdoor air pollutants altered brain function. A study of humans exposed to diesel fumes was published by Borm et al. in *Particle and Fibre Toxicology*. The EEG findings indicated a "significant" cortical stress response to the fumes.

176 *Michigan State University:* Harkema, Islam, and Pestka, "Satratoxin G."

177 *Patients with Korsakoff's:* Wilson and Stevenson, *Learning to Smell,* 168–70; also see Oliver Sacks's "The Lost Mariner," one of the case studies in *The Man Who Mistook His Wife for a Hat.*
Huntington's patients: Wilson and Stevenson, *Learning to Smell,* 170–71.

180 *Yale researchers:* Recent work is described in Carlson et al., "Odor Coding."

181 *odor perception in frogs:* University of South Carolina biologist Richard G. Vogt discusses such trends in "How Sensitive Is a Nose?"
Not all are thrilled: David J. Anderson was profiled in "Drosophila Envy," by Jeff Miller, published in the November 2006 *HHMI Bulletin.*

182 *William Frey:* See Talan, "Intranasal Delivery."

183 *Northwestern's Jay Gottfried:* Hummel and Welge-Lüssen, eds., *Taste and Smell.*

20. NO QUICK FIX

185 *smell actually* outlasts: Johnston et al., "Olfactory Ability." Griffith University scientists studied almost a thousand Australian males and females of all ages.
some hearing loss: Nagourney, "Screening." While the study, conducted at Johns Hopkins University, reconfirmed that hearing loss is prevalent in older people, researchers were surprised to find that hearing loss is more common in younger people than previously believed.

186 *A ninety-five-year-old:* This story was reported by National Public Radio in January 2009.

187 *The same is true:* The announcement of the world's first bionic eye (including the statement by the inventors that it would not lead to an implantable eye for humans) was reported on National Public Radio's *All Things Considered* in 2008.

A man named Mike May: Kurson, *Crashing Through.*

LeDoux explained that: LeDoux, *Synaptic Self.* LeDoux went on to write that "the problem of understanding this is called the binding problem" (page 193). One popular solution involves the notion of neuronal synchrony. This "simultaneous firing" is the basis for one explanation of consciousness.

188 *chemotherapy was destroying:* Burstein, "Olfactory Hallucinations."

189 *At the same time:* These issues are discussed by Matt Ridley in his book *Nature via Nurture.* Richard Doty offered some insight on shifting priorities in funding for science and treatment in olfactory medicine.

192 *the man who invented:* Svoboda, "Sniff Test."

195 *four out of ten chefs:* Hirsch discussed his informal experiment with me in an interview in 2006.

The last time: ABC's August 1, 2008, *20/20* segment entitled "Eat Ice Cream, Burgers and Pizza and Still Lose Weight?" discussed how Sprinkle Thin became Sensa and why neither product kept weight off.

21. GOING AFTER ZICAM

196 *rubbed Vicks:* Detailed information on Vicks VapoRub can be found in the papers of Henry Smith Richardson, located in the manuscripts department of the Library of the University of North Carolina at Chapel Hill.

197 *plaintiffs in the Zicam:* References to the class-action lawsuit are from various sources, including the *Washington Post.* See Boodman's "Paying Through the Nose."

198 *An Australian study:* Johnston et al., "Olfactory Ability."

199 *website called nosmell.com:* The site apparently shut down in 2007.

An online blogger: The victim told his story on the blog tvjunkie .typepad.com on December 23, 2007. The piece begins, "I thought I'd tell you a little story to help keep you from making the same mistake I did last year. That mistake . . . was Zicam."

200 *sales of Zicam products:* Matrixx Initiatives posts its balance sheet on-line.

203 *a massive dose:* Sanders, "Fear of Falling."

EPILOGUE

218 *fallen on deaf ears:* Morgenson, "Eyeshade Smelled Trouble." The *Times* business writer also reported that the Securities and Exchange Commission had begun an inquiry into the company. She related this turn of events to an incident beginning in 2002 when Matrixx filed a defamation suit against Internet posters who spoke out against Zicam products. Around the same time, an independent stock analyst and CPA had repeatedly warned investors that the company's stock value was inflated and the company itself was in jeopardy because of potential litigation from aggrieved customers. As part of the defamation suit, Matrixx barraged him with expensive subpoena requests, which caused him to have to shut down his website.